Spatial Statistics for Data Science

Spatial data is crucial to improve decision-making in a wide range of fields including environment, health, ecology, urban planning, economy, and society. **Spatial Statistics for Data Science: Theory and Practice with R** describes statistical methods, modeling approaches, and visualization techniques to analyze spatial data using R. The book provides a comprehensive overview of the varying types of spatial data, and detailed explanations of the theoretical concepts of spatial statistics, alongside fully reproducible examples which demonstrate how to simulate, describe, and analyze spatial data in various applications. Combining theory and practice, the book includes real-world data science examples such as disease risk mapping, air pollution prediction, species distribution modeling, crime mapping, and real state analyses. The book utilizes publicly available data and offers clear explanations of the R code for importing, manipulating, analyzing, and visualizing data, as well as the interpretation of the results. This ensures contents are easily accessible and fully reproducible for students, researchers, and practitioners.

Key Features:
- Describes R packages for retrieval, manipulation, and visualization of spatial data
- Offers a comprehensive overview of spatial statistical methods including spatial autocorrelation, clustering, spatial interpolation, model-based geostatistics, and spatial point processes
- Provides detailed explanations on how to fit and interpret Bayesian spatial models using the integrated nested Laplace approximation (INLA) and stochastic partial differential equation (SPDE) approaches

Paula Moraga is Professor of Statistics at King Abdullah University of Science and Technology (KAUST). She received her Master's in Biostatistics from Harvard University and her Ph.D. in Mathematics from the University of Valencia. Dr. Moraga develops innovative statistical methods and open-source software for spatial data analysis and health surveillance, including R packages for spatio-temporal modeling, detection of clusters, and travel-related spread of disease. Her work has directly informed strategic policy in reducing the burden of diseases such as malaria and cancer in several countries. Dr. Moraga has published extensively in leading journals, and serves as an Associate Editor of the *Journal of the Royal Statistical Society Series A*. She is the author of the book *Geospatial Health Data: Modeling and Visualization with R-INLA and* Shiny (Chapman & Hall/CRC). Dr. Moraga received the prestigious Letten Prize for her pioneering research in disease surveillance, and her significant contributions to the development of sustainable solutions for health and environment globally.

CHAPMAN & HALL/CRC DATA SCIENCE SERIES

Reflecting the interdisciplinary nature of the field, this book series brings together researchers, practitioners, and instructors from statistics, computer science, machine learning, and analytics. The series will publish cutting-edge research, industry applications, and textbooks in data science.

The inclusion of concrete examples, applications, and methods is highly encouraged. The scope of the series includes titles in the areas of machine learning, pattern recognition, predictive analytics, business analytics, Big Data, visualization, programming, software, learning analytics, data wrangling, interactive graphics, and reproducible research.

Recently Published Titles

Practitioner's Guide to Data Science
Hui Lin and Ming Li

Natural Language Processing in the Real World
Text Processing, Analytics, and Classification
Jyotika Singh

Telling Stories with Data
With Applications in R
Rohan Alexander

Big Data Analytics
A Guide to Data Science Practitioners Making the Transition to Big Data
Ulrich Matter

Data Science for Sensory and Consumer Scientists
Thierry Worch, Julien Delarue, Vanessa Rios De Souza and John Ennis

Data Science in Practice
Tom Alby

Introduction to NFL Analytics with R
Bradley J. Congelio

Soccer Analytics: An Introduction Using R
Clive Beggs

Spatial Statistics for Data Science: Theory and Practice with R
Paula Moraga

Research Software Engineering: Research Software Engineering
Matthias Bannert

For more information about this series, please visit: https://www.routledge.com/Chapman--HallCRC-Data-Science-Series/book-series/CHDSS

Spatial Statistics for Data Science

Theory and Practice with R

Paula Moraga

CRC Press
Taylor & Francis Group
Boca Raton London New York

CRC Press is an imprint of the
Taylor & Francis Group, an **informa** business

A CHAPMAN & HALL BOOK

Designed cover image: © Paula Moraga

First edition published 2024
by CRC Press
2385 NW Executive Center Drive, Suite 320, Boca Raton FL 33431

and by CRC Press
4 Park Square, Milton Park, Abingdon, Oxon, OX14 4RN

CRC Press is an imprint of Taylor & Francis Group, LLC

© 2024 Paula Moraga

Library of Congress Cataloging-in-Publication Data

Names: Moraga, Paula, author.
Title: Spatial statistics for data science : theory and practice with R /
Paula Moraga.
Description: First edition. | Boca Raton, FL : CRC Press, 2024. | Series:
Chapman & Hall/CRC data science series | Includes bibliographical
references and index.
Identifiers: LCCN 2023035998 (print) | LCCN 2023035999 (ebook) | ISBN
9781032633510 (hardback) | ISBN 9781032641485 (paperback) | ISBN
9781032641522 (ebook)
Subjects: LCSH: Spatial analysis (Statistics) | R (Computer program
language)
Classification: LCC QA278.2 .M684 2024 (print) | LCC QA278.2 (ebook) |
DDC 001.4/22--dc23/eng/20231023
LC record available at https://lccn.loc.gov/2023035998
LC ebook record available at https://lccn.loc.gov/2023035999

ISBN: 978-1-032-63351-0 (hbk)
ISBN: 978-1-032-64148-5 (pbk)
ISBN: 978-1-032-64152-2 (ebk)

DOI: 10.1201/9781032641522

Typeset in Latin Modern font
by KnowledgeWorks Global Ltd.

Publisher's note: This book has been prepared from camera-ready copy provided by the authors.

To Pepe, Bernar, Gonzalo, and
the memory of my beloved parents.

Contents

Preface

Spatial Statistics for Data Science: Theory and Practice with R describes statistical methods, modeling approaches, and visualization techniques to analyze spatial data using R. The book starts by providing a comprehensive overview of the types of spatial data and R packages for spatial data retrieval, manipulation, and visualization. Then, it provides a detailed explanation of the theoretical concepts of spatial statistics, along with fully reproducible examples demonstrating how to simulate, describe, and analyze areal, geostatistical, and point pattern data in various applications.

The book combines theory and practice using real-world data science examples such as disease risk mapping, air pollution prediction, species distribution modeling, crime mapping, and real state analyses. The book covers the following topics:

- Spatial data including areal, geostatistical, and point patterns
- Coordinate reference systems and geographical data storages
- R packages for retrieval, manipulation, and visualization of spatial data
- Statistical methods to simulate, describe, and analyze spatial data
- Areal data: neighborhood matrices, spatial autocorrelation, Bayesian spatial models
- Geostatistical data: Gaussian random fields, spatial interpolation, Kriging, model-based geostatistics
- Point patterns: kernel intensity estimation, clustering, log-Gaussian Cox processes
- Fitting and interpreting Bayesian spatial models using the integrated nested Laplace approximation (INLA) and stochastic partial differential equation (SPDE) approaches
- Model assessment criteria and cross-validation
- Effective communication using interactive visualizations and dashboards

The book utilizes publicly available data and offers clear explanations of the R code for importing, manipulating, analyzing, and visualizing data, as well as the interpretation of the results. This ensures contents are easily accessible and reproducible for students, researchers, and practitioners.

Audience

This book serves as a valuable resource to anyone interested in the theoretical and practical aspects of spatial statistics, with a focus on applying these methods using R. This includes statisticians, data scientists, epidemiologists, environmental scientists, geographers, urban planners, climate scientists, and professionals of government agencies looking to deepen their understanding of spatial data analysis. The book is also appropriate for students of statistics and data science, as well as other fields with a strong statistical background. The book provides readers with a solid foundation in the theory of spatial statistics, as well as practical skills for working with spatial data using R for data retrieval, manipulation, and visualization across a range of disciplines.

Prerequisites and recommended reading

Readers are assumed to have a good understanding of statistical concepts such as probability distributions, descriptive statistics, confidence intervals, hypothesis testing, and generalized linear models. The book employs the statistical software R to illustrate methods and examples. It is assumed readers have some basic understanding of R programming, including how to install and load packages, manipulate data objects, and create plots. Books that can assist readers in enhancing their R skills include Grolemund (2014) and Wickham and Grolemund (2016), which provide friendly introductions to R and data analysis with hands-on examples. Moraga (2019) describes spatial and spatio-temporal models in health and environmental applications. It also shows how to easily turn analyses into visually informative reports, dashboards, and Shiny web applications for reproducible research and communication.

Why read this book?

Spatial data arise in many fields including environment, health, ecology, agriculture, urban planning, economy, and society. The utilization of spatial data has emerged as a critical component in data science, serving as a powerful tool for governments, companies, and individuals to improve their decision-making processes. A significant example is the utilization of spatial data by statistical offices across the world to improve the assessment and surveillance of the United Nations' Sustainable Development Goals (SDGs). Spatial data are crucial to understand patterns of health outcomes and risk factors, monitor and manage natural resources, analyze demographics, design cities, preserve endangered species, and rapidly detect infectious disease outbreaks.

This book provides a comprehensive reference to spatial statistics for data science, supported by practical and fully reproducible examples across diverse fields. The statistical software R is used throughout the book, providing a wide range of packages and functions for handling spatial data in different formats, and facilitating analysis and visualization. R is available for download and use for free, making it an accessible option for researchers, educators, and practitioners. By employing the cutting-edge methods presented in the book, readers can gain valuable insights that support informed decision-making across a wide range of fields including public health, environment, and business.

Structure of the book

This book consists of four parts and an appendix.

Part I. Spatial data

The objective of the first part of the book is to present readers to the different types of spatial data, storage files of spatial data, and coordinate reference systems. This part also introduces packages that can be used to create, read, manipulate, and write spatial data in R. Additionally, it presents packages that facilitate the creation of maps, and packages that allow us to download open spatial data.

Part II. Areal data

The second part is devoted to the analysis of areal data. This type of data arise when a study region is partitioned into a finite number of areas at which outcomes are aggregated. Examples of areal data include the number of individuals with a certain disease in municipalities of a country or the average housing prices in districts of a city. This part covers key concepts such as spatial neighborhood matrices and spatial autocorrelation. It also shows how to fit and interpret Bayesian spatial models to analyze areal data. Examples of disease mapping and housing prices prediction are used to illustrate the application of these techniques.

Part III. Geostatistical data

The third part of the book is centered on geostatistical data, which refers to measurements of a spatially continuous phenomenon collected at specific locations, such as air pollution or temperature levels taken at a set of monitoring stations. This part provides an introduction to Gaussian fields and R packages used for their simulation and analysis. In addition, it presents various spatial interpolation methods including inverse distance weighted methods, Kriging, and model-based geostatistics. These methods are illustrated using several examples such as the prediction of soil metal concentrations and air pollution levels. This part also covers measures to assess the predictive performance of the interpolation methods using cross-validation techniques.

Part IV. Spatial point patterns

Spatial point patterns are countable sets of points that arise as realizations of stochastic spatial point processes within a planar region. Examples of point patterns include the locations of trees in a forest, addresses of individuals with a disease in a city, and the locations of cells in a tissue. The fourth part of the book provides an overview of techniques for analyzing point pattern data, including methods to assess the randomness of spatial point patterns, intensity estimation and clustering analysis. It also demonstrates how to formulate and fit log-Gaussian Cox process models for point pattern data, which are typically used to model environmentally driven phenomena. Examples in this part include the analysis of disease data, crime mapping, and species distribution modeling.

Appendix

Finally, the appendix provides useful resources on the R software and packages for visualization, as well as the creation of interactive dashboards and Shiny web applications to effectively communicate results to collaborators and policymakers.

Acknowledgments

R is a powerful and accessible tool for spatial statistics and data science. I am grateful to the R community, and the developers and contributors of open-source software for providing valuable resources that enable the analysis of spatial data. This book is the result of a compilation of teaching materials developed over several years. I am grateful to the students of the courses where I had the opportunity to teach for their feedback and insightful questions that helped me improve this book. Finally, I would also like to express my sincere gratitude to my research group and all my collaborators over the years for the opportunity to work with them on great problems to advance spatial statistics and data science.

Paula Moraga
KAUST

About the author

Dr. Paula Moraga (https://www.paulamoraga.com/) is an Assistant Professor of Statistics at King Abdullah University of Science and Technology (KAUST). Previously, she held academic statistics positions at Lancaster University, Harvard School of Public Health, London School of Hygiene & Tropical Medicine, Queensland University of Technology, and University of Bath. Dr. Moraga received her master's degree in Biostatistics from Harvard University, and her Ph.D. in Mathematics from the University of Valencia.

Dr. Moraga has worked in statistical research for over a decade, with a strong focus on spatial epidemiology and modeling. She develops innovative statistical methods and open-source software for spatial data analysis and health surveillance, including methods to understand geographic and temporal patterns of diseases, assess their relationship with potential risk factors, identify clusters, measure inequalities, and quickly detect outbreaks. Her work has directly informed strategic policy in reducing the burden of diseases such as malaria and cancer in several countries.

Dr. Moraga has published extensively in leading journals, and serves as an Associate Editor of the *Journal of the Royal Statistical Society Series A*. She has been invited to deliver multiple training courses on spatial statistics and the development of interactive visualization applications using R. She has also created educational materials that impact learning on a large scale, including her book *Geospatial Health Data: Modeling and Visualization with R-INLA and Shiny* (Moraga, 2019).

Dr. Moraga is the 2023 winner of the prestigious Letten Prize. Awarded by the Letten Foundation and the Young Academy of Norway, the prize recognizes young researchers' contributions to health, development, environment, and equality across all aspects of human life. Dr. Moraga received the Letten Prize for her pioneering research in disease surveillance, and her significant contributions to the development of sustainable solutions for health and environment globally.

Part I

Spatial data

1

Types of spatial data

Spatial data are used across a wide range of fields to support decision-making, including environment, public health, ecology, agriculture, urban planning, economy, and society. These data arise from various sources and are available in multiple formats (Moraga and Baker, 2022). For instance, remote sensing data such as land use and environmental phenomena can be obtained through satellites orbiting the Earth and other distance-capturing platforms. Monitoring stations located at specific sites provide detailed information on various environmental and climatic variables such as temperature, rainfall, and air pollution. Surveys are employed to gather data on different social, economic, and health-related topics. Spatial data can also be derived from mobile phone usage and social media which can provide information on the location and activities of individuals.

Spatial data can be thought of as resulting from observations of a stochastic process

$$\{Z(\boldsymbol{s}) : \boldsymbol{s} \in D \subset \mathbb{R}^d\},$$

where D is a set of $\mathbb{R}^d$, $d = 2$, and $Z(\boldsymbol{s})$ denotes the attribute we observe at $\boldsymbol{s}$. Three types of spatial data are distinguished through the characteristics of the domain D, namely, areal (or lattice) data, geostatistical data, and point patterns (Cressie, 1993). Below we describe each of the data types, and give examples of these data in different settings.

1.1   Areal data

In areal or lattice data, the domain D is a fixed countable collection of (regular or irregular) areal units at which variables are observed. Areal data usually arise when the number of events corresponding to some variable of interest are aggregated in areas. For example, in spatial epidemiology, locations of individuals with a given disease are often aggregated in administrative areas. These data can be analyzed to understand geographic patterns and identify factors of disease risk, taking into account the neighborhood configuration and other factors known to affect disease risk (Moraga, 2018a). Areal data may also arise in remote sensing applications where satellites provide information on a number of variables such as temperature, precipitation, and vegetation indices at cells of a regular grid that covers the study region.

Examples

Figure 1.1 shows the number of sudden infant deaths in each of the counties of North Carolina, USA, in 1974 from the **sf** package (Pebesma, 2022a).

```
library(sf)
library(mapview)
d <- st_read(system.file("shape/nc.shp", package = "sf"),
             quiet = TRUE)
mapview(d, zcol = "SID74")
```

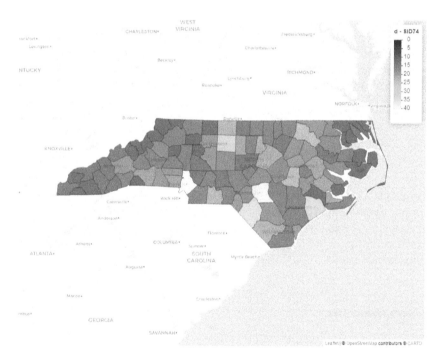

FIGURE 1.1: Example of areal data. Number of sudden infant deaths in counties of North Carolina, USA, in 1974.

The map in Figure 1.2 depicts household income in $1000 USD in neighborhoods in Columbus, Ohio, in 1980 contained in the **spData** package (Bivand et al., 2022).

```
library(spData)
library(ggplot2)
d <- st_read(system.file("shapes/columbus.shp",
                         package = "spData"), quiet = TRUE)
ggplot(d) + geom_sf(aes(fill = INC))
```

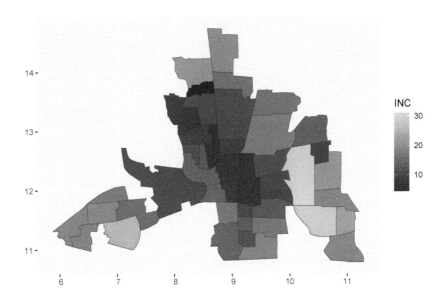

FIGURE 1.2: Example of areal data. Household income in $1000 USD in neighborhoods in Columbus, Ohio, in 1980.

Figure 1.3 shows elevation at raster grid cells covering Luxembourg from **terra** (Hijmans, 2022). In this case, areas are all of the same size equal to the cells of a raster grid.

```
library(terra)
d <- rast(system.file("ex/elev.tif", package = "terra"))
plot(d)
```

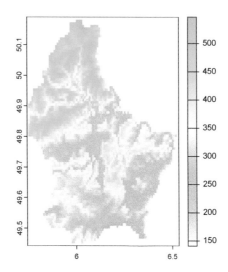

FIGURE 1.3: Example of areal data. Elevation at raster grid cells covering Luxembourg.

1.2   Geostatistical data

In geostatistical data, D is a continuous fixed subset of $\mathbb{R}^d$. The spatial index s varies continuously in space and therefore $Z(s)$ can be observed everywhere within D. Usually, we use data $\{Z(s_1), \ldots, Z(s_n)\}$ observed at known spatial locations $\{s_1, \ldots, s_n\}$ to predict the values of the variable of interest at unsampled locations. For example, we can use air pollution measurements at a number of monitoring stations to predict air pollution at other locations taking into account spatial autocorrelation and other factors that are known to predict the outcome of interest (Cameletti et al., 2013).

Examples

Figure 1.4 shows topsoil lead concentrations (mg per kg of soil) at several locations sampled in a flood plain of the river Meuse, near Stein, The Netherlands, obtained from the **sp** package (Pebesma and Bivand, 2022).

```
library(sp)
library(sf)
library(mapview)
```

```
data(meuse)
meuse <- st_as_sf(meuse, coords = c("x", "y"), crs = 28992)
mapview(meuse, zcol = "lead",  map.types = "CartoDB.Voyager")
```

FIGURE 1.4: Example of geostatistical data. Topsoil lead concentrations at locations sampled in a flood plain of the river Meuse, The Netherlands.

The map in Figure 1.5 shows the price per square meter (Euros per square meter) of a specific set of apartments in Athens, Greece, in 2017 from **spData** (Bivand et al., 2022).

```
library(spData)
mapview(properties, zcol = "prpsqm")
```

Figure 1.6 shows malaria prevalence at specific locations in Zimbabwe from the **malariaAtlas** package (Pfeffer et al., 2020). Prevalence is calculated as the number of individuals positive for malaria divided by the number of examined individuals at each of the locations.

```
library(malariaAtlas)
d <- getPR(country = "Zimbabwe", species = "BOTH")
ggplot2::autoplot(d)
```

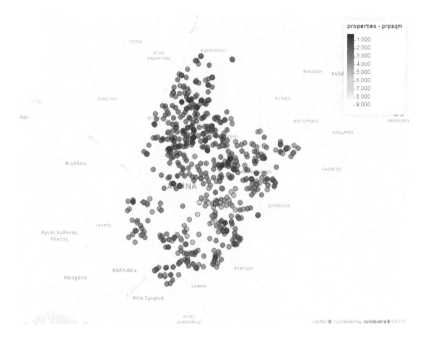

FIGURE 1.5: Example of geostatistical data. Price per square meter of a set of apartments in Athens, Greece, in 2017.

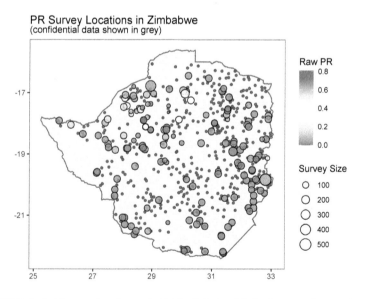

FIGURE 1.6: Example of geostatistical data. Malaria prevalence at specific locations in Zimbabwe.

1.3   Point patterns

Finally, in point patterns, the domain D is random. Its index set gives the locations of random events of the spatial point pattern, and $Z(s)$ may be equal to 1 $\forall s \in D$, indicating occurrence of the event, or random, giving some additional information.

Point patterns arise when the variable to be analyzed corresponds to the location of events. For example, patterns may include the locations of fires in a forest (González and Moraga, 2022) or the residential addresses of people with a disease (Moraga and Montes, 2011). Often, we are interested in understanding the underlying spatial process that originates the point pattern, and assessing whether the spatial pattern exhibits randomness, clustering, or regularity.

Examples

An example of spatial point pattern is the fires in Castilla-La Mancha, Spain, between 1998 and 2007 contained in the `clmfires` data of the **spatstat** package (Baddeley et al., 2022). Data `clmfires` is a marked point pattern containing information of each fire. Figure 1.7 depicts the location of the fires without the mark.

```
library(spatstat)
plot(clmfires, use.marks = FALSE, pch = ".")
```

This figure also shows the positions of cell nuclei in a histological section of a tissue from a lymphoma in the kidney of a hamster from **spatstat**. The nuclei are classified as either "pyknotic" (corresponding to dying cells) or "dividing" (corresponding to cells arrested in the act of dividing).

```
library(spatstat)
plot(hamster)
```

Figure 1.8 shows the spatial locations of 761 cases of primary biliary cirrhosis and 30210 controls representing at-risk population in north-eastern England collected between 1987 and 1994. This information is contained in the `pbc` data from the **sparr** package (Davies and Marshall, 2023).

clmfires

hamster

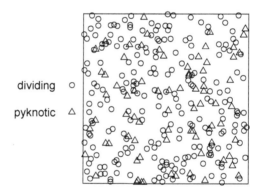

FIGURE 1.7: Examples of point patterns. Top: Locations of fires in Castilla-La Mancha, Spain, between 1998 and 2007. Bottom: Locations and types of cells in a tissue.

```
library(sparr)
data(pbc)
plot(unmark(pbc[which(pbc$marks == "case"), ]), main = "cases")
axis(1)
axis(2)
title(xlab = "Easting", ylab = "Northing")
```

```
plot(unmark(pbc[which(pbc$marks == "control"), ]),
     pch = 3, main = "controls")
axis(1)
axis(2)
title(xlab = "Easting", ylab = "Northing")
```

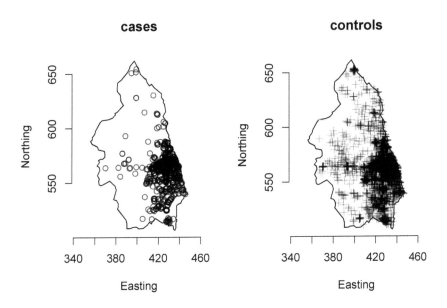

FIGURE 1.8: Example of point pattern. Locations of cases and controls of primary biliary cirrhosis in north-eastern England between 1987 and 1994.

1.4 Spatio-temporal data

Spatio-temporal data arise when information is both spatially and temporally referenced. Thus, we can consider spatial data as temporal aggregations or temporal snapshots of a spatio-temporal process. Examples of spatio-temporal data include the number of car accidents in each of the US states in each of the months of 2020 (areal data), air pollution levels measured at 100 monitoring stations located in Germany each hour of a given day (geostatistical data), and the locations of earthquakes occurring in the world in each of the years from 2000 to 2020 (point pattern). Figure 1.9 shows a spatio-temporal dataset

representing the population of the counties of Ohio, USA, from 1968 to 1988
obtained from the **SpatialEpiApp** package (Moraga, 2017).

```r
# devtools::instzall_github("Paula-Moraga/SpatialEpiApp")
library(SpatialEpiApp)
library(sf)
library(ggplot2)
library(viridis)

# map
f <- file.path("SpatialEpiApp/data/Ohio/fe_2007_39_county/",
               "fe_2007_39_county.shp")
pathshp <- system.file(f, package = "SpatialEpiApp")
map <- st_read(pathshp, quiet = TRUE)

# datu
namecsv <- "SpatialEpiApp/data/Ohio/dataohiocomplete.csv"
d <- read.csv(system.file(namecsv, package = "SpatialEpiApp"))

# data are disaggregated by gender and race
# aggregate to get population in each county and year
d <- aggregate(x = d$n, by = list(county = d$NAME, year = d$year),
               FUN = sum)
names(d) <- c("county", "year", "population")

# join map and data
mapst <- dplyr::left_join(map, d, by = c("NAME" = "county"))

# map population by year
# facet_wrap() splits data into subsets and create multiple plots
ggplot(mapst, aes(fill = log(population))) + geom_sf() +
  facet_wrap(~ year, ncol = 7) +
  scale_fill_viridis("log(population)") +
  theme(axis.text.x = element_blank(),
        axis.text.y = element_blank(),
        axis.ticks = element_blank())
```

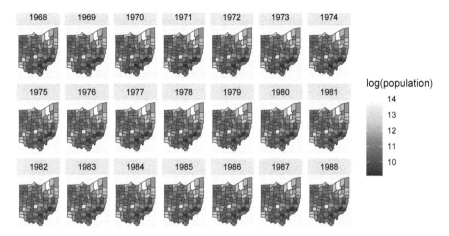

FIGURE 1.9: Example of spatio-temporal data. Population of the counties of Ohio, USA, from 1968 to 1988.

1.5 Spatial functional data

Spatial functional data arise when the three types of spatial data (areal, geostatistical, and point patterns) are combined with random functions. Thus, a spatial functional process can be defined as

$$\{\chi_s : s \in D \subset \mathbb{R}^d\},$$

where χ_s is a functional random variable taking values in an infinite dimensional space observed at s in the spatial domain D. Typically, χ_s is a real function from $[a, b] \subset \mathbb{R}$ to $\mathbb{R}$.

The spatial domain D can be fixed or random and allows us to classify spatial functional data as functional areal data when the functions correspond to areas, functional geostatistical data when functions are observed at a fixed subset of locations, and functional point patterns when functions are observed at each of the locations of a point process.

Example

The example below shows functional geostatistical data from the **geoFourierFDA** package (Sassi, 2021) representing the daily temperature averaged over 30 years at 35 Canadian weather stations, $\{\chi_{s_i} : i = 1, \ldots, 35\}$. Figure 1.10 shows the locations of the Canadian weather stations, and Figure 1.11 the daily temperature measured in each of the stations. One of the objectives when analyzing this data could be the prediction of the daily temperature function $\chi_{s_0} : [0, 365) \to \mathbb{R}$ at one specific unsampled location s_0 in Canada.

```r
library(sf)
library(geoFourierFDA)
library(rnaturalearth)

# Map Canada
map <- rnaturalearth::ne_states("Canada", returnclass = "sf")

# Coordinates of stations
d <- data.frame(canada$m_coord)
d$location <- attr(canada$m_coord, "dimnames")[[1]]
d <- st_as_sf(d, coords = c("W.longitude", "N.latitude"))
st_crs(d) <- 4326

# Plot Canada map and location of stations
ggplot(map) + geom_sf() + geom_sf(data = d, size = 6) +
  geom_sf_label(data = d, aes(label = location), nudge_y = 2)


# Temperature of each station over time
d <- data.frame(canada$m_data)
d$time <- 1:nrow(d)

# Pivot data d from wide to long
# cols: columns to pivot in longer format
# names_to: name of new column with column names of original data
# values_to: name of new column with values of original data
df <- tidyr::pivot_longer(data = d,
cols = names(d)[-which(names(d) == "time")],
names_to = "variable", values_to = "value")

# Plot temperature of each station over time
ggplot(df, aes(x = time, y = value)) +
  geom_line(aes(color = variable))
```

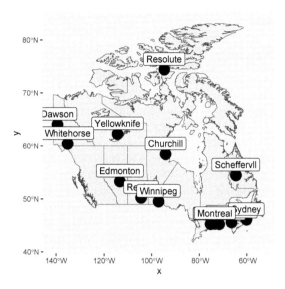

FIGURE 1.10: Locations of Canadian weather stations where daily temperature is measured.

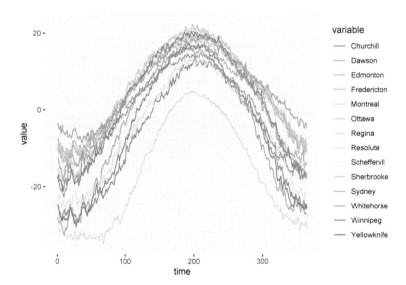

FIGURE 1.11: Example of spatial functional data. Daily temperature averaged over 30 years measured at 35 Canadian weather stations.

1.6   Mobility data

Besides the three classical types of spatial data (i.e., areal, geostatistical, and point patterns), we can also consider other spatial data such as flows containing the number of individuals or other elements moving between locations (Mahmood et al., 2022). Here, we see an example of flows data from the **epiflows** package (Piatkowski et al., 2018; Moraga et al., 2019). This package allows us to predict and visualize the spread of infectious diseases based on flows between geographical locations. The package contains the `Brazil_epiflows` data with the number of travelers between Brazilian states and other locations. We can use this data to create an `epiflows` object called `ef` that allows us to use the prediction and visualization functions. Then, we can visualize the population flows with `vis_epiflows(ef)` using a dynamic network, and `map_epiflows(ef)` using an interactive map.

```
library("epiflows")
data("Brazil_epiflows")

loc <- merge(x = YF_locations, y = YF_coordinates,
by.x = "location_code", by.y = "id", sort = FALSE)

ef <- make_epiflows(flows = YF_flows, locations = loc,
                    coordinates = c("lon", "lat"),
                    pop_size = "location_population",
                    duration_stay = "length_of_stay",
                    num_cases = "num_cases_time_window",
                    first_date = "first_date_cases",
                    last_date = "last_date_cases")
```

2

Spatial data in R

Spatial data can be represented using vector and raster data. Vector data is used to display points, lines, and polygons, and possibly associated information. Vector data may represent, for example, locations of monitoring stations, road networks, or municipalities of a country. Raster data are regular grids with cells of equal size that are used to store values of spatially continuous phenomena, such as elevation, temperature, or air pollution values.

The **sf** (Pebesma, 2022a) and **terra** (Hijmans, 2022) packages are the main packages that allow us to manipulate and analyze spatial data in R. In this chapter, we introduce these packages, spatial data storage files, and coordinate reference systems. Finally, we give an overview of old spatial packages that were widely used but are not longer maintained.

2.1 Vector data

The **sf** package allows us to work with vector data which is used to represent points, lines, and polygons (Figure 2.1). Vector data can be used, for example, to represent locations of hospitals or monitoring stations as points, roads or rivers as lines, and municipalities or districts of a country as polygons. Moreover, these data can also have associated information such as temperature values measured at monitoring stations or number of people living in municipalities.

Before the **sf** package was developed, the **sp** package (Pebesma and Bivand, 2022), which is no longer maintained, was used to work with vector spatial data. The **terra** package presented in the following sections is mainly used to work with rasters and also has functionality to work with vector data.

2.1.1 Shapefile

Vector data are often represented using a data storage format called shapefile. Note that a shapefile is not a single file but a collection of related files. A shapefile has three mandatory files, namely, .shp which contains the geometry data, .shx which is a positional index of the geometry data that allows to

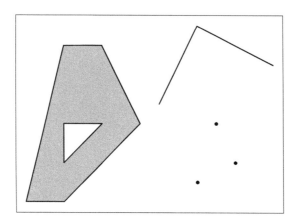

FIGURE 2.1: Examples of vector data (polygon, line, and points).

seek forward and backward the `.shp` file, and `.dbf` which stores the attributes for each shape. Other files that may form a shapefile include `.prj` which is a plain text file describing the projection, `.sbn` and `.sbx` which are spatial indices of the geometry data, and `.shp.xml` which contains spatial metadata in XML format. Therefore, when working with a shapefile, it is important to obtain all files that compose the shapefile and not only the `.shp` file with the geometry data.

The `st_read()` function of the **sf** package can be used to read a shapefile. Here, we read the shapefile of the counties of North Carolina, USA, contained in the **sf** package. First, we use `system.file()` passing the name of the directory (`"shape/nc.shp"`) and the name of the package (`"sf"`) to identify the path of shapefile.

```
library(sf)
pathshp <- system.file("shape/nc.shp", package = "sf")
```

We can examine the **shape** directory and see that it contains the following files corresponding to the North Carolina shapefile.

```
shape
└── nc.shp
└── nc.shx
└── nc.dbf
└── nc.prj
```

Then, we read the shapefile with `st_read()` passing the name to read the shapefile. We set `quiet = TRUE` to suppress information on name, driver, size, and spatial reference.

```
map <- st_read(pathshp, quiet = TRUE)
class(map)


[1] "sf"          "data.frame"


head(map)


Simple feature collection with 6 features and 14 fields
Geometry type: MULTIPOLYGON
Dimension:     XY
Bounding box:  xmin: -81.74 ymin: 36.07 xmax: -75.77 ymax: 36.59
Geodetic CRS:  NAD27
    AREA PERIMETER CNTY_  CNTY_ID          NAME  FIPS
1 0.114     1.442  1825     1825          Ashe 37009
2 0.061     1.231  1827     1827     Alleghany 37005
3 0.143     1.630  1828     1828         Surry 37171
4 0.070     2.968  1831     1831      Currituck 37053
5 0.153     2.206  1832     1832 Northampton 37131
6 0.097     1.670  1833     1833     Hertford 37091
  FIPSNO CRESS_ID BIR74 SID74 NWBIR74 BIR79 SID79
1  37009        5  1091     1      10  1364     0
2  37005        3   487     0      10   542     3
3  37171       86  3188     5     208  3616     6
4  37053       27   508     1     123   830     2
5  37131       66  1421     9    1066  1606     3
6  37091       46  1452     7     954  1838     5
  NWBIR79                       geometry
1      19 MULTIPOLYGON (((-81.47 36.2...
2      12 MULTIPOLYGON (((-81.24 36.3...
3     260 MULTIPOLYGON (((-80.46 36.2...
4     145 MULTIPOLYGON (((-76.01 36.3...
5    1197 MULTIPOLYGON (((-77.22 36.2...
6    1237 MULTIPOLYGON (((-76.75 36.2...
```

Figure 2.2 shows the first attribute of the map.

```
plot(map[1]) # plot first attribute
```

AREA

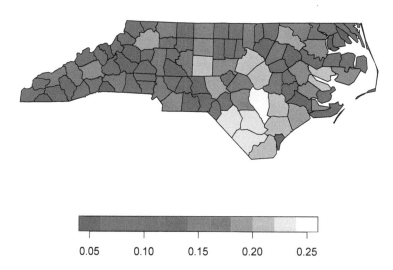

FIGURE 2.2: Map of the first attribute of the 'sf' object representing the counties of North Carolina, USA.

2.2 Raster data

Raster data (also referred to as grid data) is a spatial data structure that divides the region of study into rectangles of the same size called cells or pixels, and that can store one or more values for each of these cells (Figure 2.3). Raster data is used to represent spatially continuous phenomena, such as elevation, temperature, or air pollution values.

In R, **terra** is the main package to work with raster data, which also has functionality to work with vector data. Before **terra** was developed, the **raster** package (Hijmans, 2023) was used to analyze raster data. **terra** is very similar to **raster** but can do more and is faster. The **stars** package (Pebesma, 2022b) can also be used to analyze raster data as well as spatial data cubes which are arrays with one or more spatial dimensions.

2.2.1 GeoTIFF

Raster data often come in GeoTIFF format which has extension `.tif`. Here, we use the `terra::rast()` function to read the `elev.tif` file of the **terra** package that represents elevation in Luxembourg (Figure 2.3).

```
library(terra)
pathraster <- system.file("ex/elev.tif", package = "terra")
r <- terra::rast(pathraster)
r
plot(r)
```

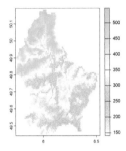

FIGURE 2.3: Left: Example of raster data with cells colored according to their values. Right: Map of raster data representing the elevation of Luxembourg.

Another format commonly used to store raster data is netCDF, which stands for network Common Data Form. R provides functionality to read, write, and manipulate netCDF files through packages such as **ncdf4** (Pierce, 2023).

2.3 Coordinate Reference Systems

The coordinate reference system (CRS) of spatial data specifies the origin and the unit of measurement of the spatial coordinates. CRSs are important for spatial data manipulation, analysis and visualization, and permit to deal with multiple data by transforming them to a common CRS. Locations on the Earth can be referenced using unprojected (also called geographic) or projected CRSs. The unprojected or geographic CRS uses latitude and longitude values to represent locations on the Earth's three-dimensional ellipsoid surface. A projected CRS uses Cartesian coordinates to reference a location on a two-dimensional representation of the Earth.

2.3.1 Geographic CRS

In a geographic CRS, latitude and longitude values are used to identify locations on the Earth's three-dimensional ellipsoid surface. Latitude values measure the angles north or south of the equator (0 degrees) and range from –90 degrees at the south pole to 90 degrees at the north pole. Longitude values measure the angles west or east of the prime meridian. Longitude values range from –180 degrees when running west to 180 degrees when running east (Figure 2.4).

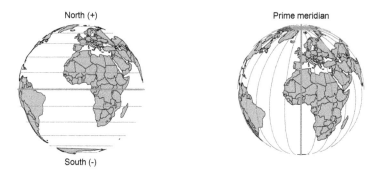

FIGURE 2.4: Parallels (left) and meridians (right).

Latitude and longitude coordinates may be expressed in degrees, minutes, and seconds, or in decimal degrees. In decimal form, northern latitudes are positive, and southern latitudes are negative. Also, eastern longitudes are positive, and western longitudes are negative. For example, the location of New York City, USA, can be given by geographic coordinates as follows:

	Latitude	Longitude
Degrees, Minutes, and Seconds	40° 43′ 50.1960″ North	73° 56′ 6.8712″ West
Decimal degrees (North/South and West/East)	40.730610° North	73.935242° West
Decimal degrees (Positive/Negative)	40.730610	–73.935242

Note here that one degree is 60 minutes, 1 minute is 60 seconds, and one degree is 3600 seconds. Note also that one degree of longitude at the equator and one degree of latitude anywhere correspond to 111.32 kilometers approximately. This means that 1 minute is equal to $111.32/60 = 1.85$ kilometers approximately. A degree of longitude is widest at the equator, and this distance shrinks as moving north or south toward the poles.

2.3.2 Projected CRS

Projected CRSs use Cartesian coordinates to reference a location on a two-dimensional representation of the Earth. All projections produce distortion of the Earth's surface in some fashion, and cannot simultaneously preserve all area, direction, shape, and distance properties. For example, Figure 2.5 shows world maps using two different projections, namely, Mercator and Robinson projections.

FIGURE 2.5: World maps with Mercator (left) and Robinson (right) projections.

A common projection is the Universal Transverse Mercator (UTM) projection[1]. This projection is conformal, meaning that it preserves angles and therefore shapes across small regions. However, it distorts distances and areas. In the UTM projection, a location is given by the zone number (60 zones), hemisphere (north or south), and Easting and Northing coordinates in the zone in meters. Eastings and Northings are referenced from the central meridian and equator, respectively, of each zone.

2.3.3 EPSG codes

Most common CRSs can be specified by providing their EPSG (European Petroleum Survey Group) codes or their Proj4 strings. Common spatial projections can be found at https://spatialreference.org/ref/. Details of a given projection can be inspected using the **st_crs()** function of the **sf** package. For example, the EPSG code 4326 refers to the WGS84 longitude/latitude projection.

```
st_crs("EPSG:4326")$Name
```

```
[1] "WGS 84"
```

```
st_crs("EPSG:4326")$proj4string
```

```
[1] "+proj=longlat +datum=WGS84 +no_defs"
```

---

[1]https://en.wikipedia.org/wiki/Universal_Transverse_Mercator_coordinate_system

```
st_crs("EPSG:4326")$epsg
```

```
[1] 4326
```

2.3.4   Transforming CRS with sf and terra

Functions `sf::st_crs()` and `terra::crs()` allow us to get the CRS of spatial data. These functions also allow us to set a CRS to spatial data by using `st_crs(x) <- value` if x is a **sf** object, and `crs(r) <- value` if r is a raster. Notice that setting a CRS does not transform the data, it just changes the CRS label. We may want to set a CRS to data that does not come with CRS, and the CRS should be what it is, not what we would like it to be. We use `sf::st_transform()` and `terra::project()` to transform the **sf** or **raster** data, respectively, to a new CRS.

For **sf** data, we can read and get the CRS, and transform the data to a new CRS as follows:

```
library(sf)
pathshp <- system.file("shape/nc.shp", package = "sf")
map <- st_read(pathshp, quiet = TRUE)

# Get CRS
# st_crs(map)
# Transform CRS
map2 <- st_transform(map, crs = "EPSG:4326")
# Get CRS
# st_crs(map2)
```

We can use **terra** to read and get the CRS, and transform the data to a new CRS of a raster as follows:

```
library(terra)
pathraster <- system.file("ex/elev.tif", package = "terra")
r <- rast(pathraster)

# Get CRS
# crs(r)
# Transform CRS
r2 <- terra::project(r, "EPSG:2169")
# Get CRS
# crs(r2)
```

Alternatively, as we may want transformed data that exactly lines up with other raster data we are using, we can project using an existing raster with the geometry we wish. For example,

```
# x is existing raster
# r is raster we project
r2 <- terra::project(r, x)
```

2.4 Old spatial packages

Before the **sf** package was developed, the **sp** package was used to represent and work with vector spatial data. **sp** as well as the **rgdal** (Bivand et al., 2023), **rgeos** (Bivand and Rundel, 2022) and **maptools** (Bivand and Lewin-Koh, 2022) packages are no longer maintained and will retire. Using old packages, the `rgdal::readOGR()` function can be used to read a file. Data can be accessed with `sp_object@data`, and the `sp::spplot()` function can be used to plot sp spatial objects.

```
library(sf)
library(sp)
library(rgdal)
pathshp <- system.file("shape/nc.shp", package = "sf")
sp_object <- rgdal::readOGR(pathshp, verbose = FALSE)
class(sp_object)
```

```
[1] "SpatialPolygonsDataFrame"
attr(,"package")
[1] "sp"
```

The `st_as_sf()` function of **sf** can be used to transform a sp object to a sf object (`st_as_sf(sp_object)`). Also, a **sf** object can be transformed to a sp object with `as(sf_object, "Spatial")`.

3

The **sf** package for spatial vector data

3.1 The sf package

The **sf** package (Pebesma, 2022a) can be used to represent and work with spatial vector data including points, polygons, and lines, and their associated information. The **sf** package uses **sf** objects that are extensions of data frames containing a collection of simple features or spatial objects with possibly associated data.

We can read a **sf** object with the `st_read()` function of **sf**. For example, here we read the **nc** shapefile of **sf** which contains the counties of North Carolina, USA, as well as their name, number of births, and number of sudden infant deaths in 1974 and 1979.

```
library(sf)
pathshp <- system.file("shape/nc.shp", package = "sf")
nc <- st_read(pathshp, quiet = TRUE)
class(nc)
```

```
[1] "sf"         "data.frame"
```

The **sf** object **nc** is a `data.frame` containing a collection with 100 simple features (rows) and 6 attributes (columns) plus a list-column with the geometry of each feature. A **sf** object contains the following objects of class **sf**, **sfc** and **sfg**:

- **sf** (simple feature): each row of the `data.frame` is a single simple feature consisting of attributes and geometry.
- **sfc** (simple feature geometry list-column): the `geometry` column of the `data.frame` is a list-column of class **sfc** with the geometry of each simple feature.
- **sfg** (simple feature geometry): each of the rows of the **sfc** list-column corresponds to the simple feature geometry (**sfg**) of a single simple feature.

We can see and plot the information of the sf object as follows (Figure 3.1):

```
print(nc)


Simple feature collection with 100 features and 14 fields
Geometry type: MULTIPOLYGON
Dimension:     XY
Bounding box:  xmin: -84.32 ymin: 33.88 xmax: -75.46 ymax: 36.59
Geodetic CRS:  NAD27
First 10 features:
    AREA PERIMETER CNTY_ CNTY_ID        NAME  FIPS
1  0.114     1.442  1825    1825        Ashe 37009
2  0.061     1.231  1827    1827   Alleghany 37005
3  0.143     1.630  1828    1828       Surry 37171
4  0.070     2.968  1831    1831   Currituck 37053
5  0.153     2.206  1832    1832 Northampton 37131
6  0.097     1.670  1833    1833    Hertford 37091
7  0.062     1.547  1834    1834      Camden 37029
8  0.091     1.284  1835    1835       Gates 37073
9  0.118     1.421  1836    1836      Warren 37185
10 0.124     1.428  1837    1837      Stokes 37169
   FIPSNO CRESS_ID BIR74 SID74 NWBIR74 BIR79 SID79
1   37009        5  1091     1      10  1364     0
2   37005        3   487     0      10   542     3
3   37171       86  3188     5     208  3616     6
4   37053       27   508     1     123   830     2
5   37131       66  1421     9    1066  1606     3
6   37091       46  1452     7     954  1838     5
7   37029       15   286     0     115   350     2
8   37073       37   420     0     254   594     2
9   37185       93   968     4     748  1190     2
10  37169       85  1612     1     160  2038     5
   NWBIR79                       geometry
1       19 MULTIPOLYGON (((-81.47 36.2...
2       12 MULTIPOLYGON (((-81.24 36.3...
3      260 MULTIPOLYGON (((-80.46 36.2...
4      145 MULTIPOLYGON (((-76.01 36.3...
5     1197 MULTIPOLYGON (((-77.22 36.2...
6     1237 MULTIPOLYGON (((-76.75 36.2...
7      139 MULTIPOLYGON (((-76.01 36.3...
8      371 MULTIPOLYGON (((-76.56 36.3...
9      844 MULTIPOLYGON (((-78.31 36.2...
10     176 MULTIPOLYGON (((-80.03 36.2...
```

```
plot(nc)
```

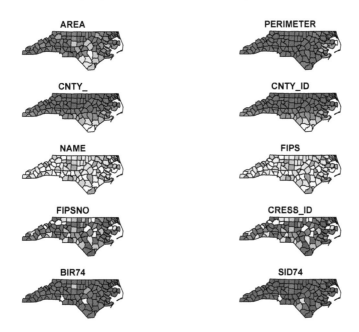

FIGURE 3.1: sf object representing the counties of North Carolina, USA, and associated information.

We can subset feature sets by using the square bracket notation and use the drop argument to drop geometries.

```
nc[1, ] # first row
nc[nc$NAME == "Ashe", ] # row with NAME "Ashe"
nc[1, "NWBIR74"] # first row, column with name NWBIR74
nc[1, "NWBIR74", drop = TRUE] # drop geometry
```

The st_geometry() function can be used to retrieve the simple feature geometry list-column (sfc).

```
# Geometries printed in abbreviated form
st_geometry(nc)
# View complete geometry by selecting one
st_geometry(nc)[[1]]
```

3.2   Creating a sf object

We can use the `st_sf()` function to create a `sf` object by providing two elements, namely, a `data.frame` with the attributes of each feature, and a simple feature geometry list-column `sfc` containing simple feature geometries `sfg`. In more detail, we create simple feature geometries `sfg` and use the `st_sfc()` function to create a simple feature geometry list-column `sfc` with them. Then, we use `st_sf()` to put the `data.frame` with the attributes and the simple feature geometry list-column `sfc` together.

Simple feature geometries `sfg` objects can be, for example, of type `POINT` (single point), `MULTIPOINT` (set of points) or `POLYGON` (polygon), and can be created with `st_point()`, `st_multipoint()` and `st_polygon()`, respectively. Here, we create a `sf` object containing two single points, a set of points, and a polygon, with one attribute. First, we create the simple feature geometry objects (`sfg`) of type `POINT`, `MULTIPOINT`, and `POLYGON`. Then, we use `st_sfc()` to create a simple feature geometry list-column `sfc` with the `sfg` objects. Finally, we use `st_sf()` to put the `data.frame` with the attribute and the simple feature geometry list-column `sfc` together. Figure 3.2 shows the resulting `sf` object plotted with **ggplot2**.

```
# Single point (point as a vector)
p1_sfg <- st_point(c(2, 2))
p2_sfg <- st_point(c(2.5, 3))

# Set of points (points as a matrix)
p <- rbind(c(6, 2), c(6.1, 2.6), c(6.8, 2.5),
           c(6.2, 1.5), c(6.8, 1.8))
mp_sfg <- st_multipoint(p)

# Polygon. Sequence of points that form a closed,
# non-self intersecting ring.
# The first ring denotes the exterior ring,
# zero or more subsequent rings denote holes in the exterior ring
p1 <- rbind(c(10, 0), c(11, 0), c(13, 2),
            c(12, 4), c(11, 4), c(10, 0))
p2 <- rbind(c(11, 1), c(11, 2), c(12, 2), c(11, 1))
pol_sfg <- st_polygon(list(p1, p2))

# Create sf object
p_sfc <- st_sfc(p1_sfg, p2_sfg, mp_sfg, pol_sfg)
df <- data.frame(v1 = c("A", "B", "C", "D"))
p_sf <- st_sf(df, geometry = p_sfc)
```

```
# Plot single points, set of points and polygon
library(ggplot2)
ggplot(p_sf) + geom_sf(aes(col = v1), size = 3) + theme_bw()
```

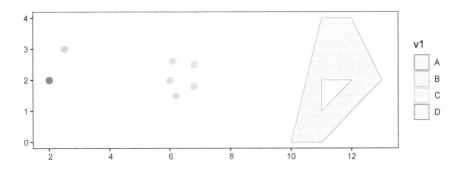

FIGURE 3.2: sf object representing two single points, a set of points, and a polygon, with one attribute.

3.3 st_*() functions

Common functions to manipulate sf objects include the following:

- st_read() reads a sf object,
- st_write() writes a sf object,
- st_crs() gets or sets a new coordinate reference system (CRS),
- st_transform() transforms data to a new CRS,
- st_intersection() intersects sf objects,
- st_union() combines several sf objects into one,
- st_simplify() simplifies a sf object,
- st_coordinates() retrieves coordinates of a sf object,
- st_as_sf() converts a foreign object to a sf object.

For example, we can read a sf object as follows:

```
library(sf)
library(ggplot2)
map <- read_sf(system.file("shape/nc.shp", package = "sf"))
```

We can inspect the first rows of the sf object map with head().

```
head(map)
```

```
Simple feature collection with 6 features and 14 fields
Geometry type: MULTIPOLYGON
Dimension:     XY
Bounding box:  xmin: -81.74 ymin: 36.07 xmax: -75.77 ymax: 36.59
Geodetic CRS:  NAD27
# A tibble: 6 x 15
   AREA PERIMETER CNTY_ CNTY_ID NAME        FIPS  FIPSNO
  <dbl>     <dbl> <dbl>   <dbl> <chr>       <chr>  <dbl>
1 0.114      1.44  1825    1825 Ashe        37009  37009
2 0.061      1.23  1827    1827 Alleghany   37005  37005
3 0.143      1.63  1828    1828 Surry       37171  37171
4 0.07       2.97  1831    1831 Currituck   37053  37053
5 0.153      2.21  1832    1832 Northampt~  37131  37131
6 0.097      1.67  1833    1833 Hertford    37091  37091
# i 8 more variables: CRESS_ID <int>, BIR74 <dbl>,
#   SID74 <dbl>, NWBIR74 <dbl>, BIR79 <dbl>,
#   SID79 <dbl>, NWBIR79 <dbl>,
#   geometry <MULTIPOLYGON [°]>
```

We can delete some of the polygons by taking a subset of the rows of `map`. We can use `st_union()` with argument `by_feature = FALSE` to combine all geometries together. The boundaries of a map can be simplified with the `st_simplify()` function (Figure 3.3).

```
# Delete polygon
map <- map[-which(map$FIPS %in% c("37125", "37051")), ]
ggplot(map) + geom_sf(aes(fill = SID79))

# Combine geometries
ggplot(st_union(map, by_feature = FALSE) %>% st_sf()) + geom_sf()

# Simplify
ggplot(st_simplify(map, dTolerance = 10000)) + geom_sf()
```

3.4   Transforming point data to a sf object

The `st_as_sf()` function allows us to convert a foreign object to a `sf` object. For example, we can have a data frame containing the coordinates of a number of locations and attributes that we wish to turn into a `sf` object. To do that,

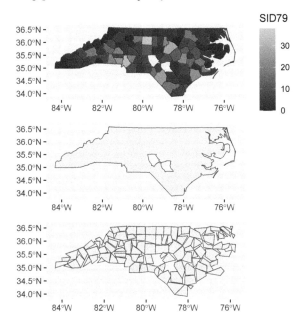

FIGURE 3.3: sf object obtained by deleting some of its polygons (top), combining polygons (middle), and simplifying polygons (bottom).

we use the st_as_sf() function passing the object we wish to convert and specifying in argument coords the name of the columns that contain the point coordinates. For example, here we use st_as_sf() to turn a data frame containing coordinates long and lat and two variables place and value to a sf object. Then, we use st_crs() to set the coordinate reference system given by the EPSG code 4326 to represent longitude and latitude coordinates. Figure 3.4 shows the plot of the sf object obtained with **mapview**.

```
library(sf)
library(mapview)

d <- data.frame(
place = c("London", "Paris", "Madrid", "Rome"),
long = c(-0.118092, 2.349014, -3.703339, 12.496366),
lat = c(51.509865, 48.864716, 40.416729, 41.902782),
value = c(200, 300, 400, 500))
class(d)
```

```
[1] "data.frame"
```

```
dsf <- st_as_sf(d, coords = c("long", "lat"))
st_crs(dsf) <- 4326
class(dsf)
```

```
[1] "sf"           "data.frame"
```

```
mapview(dsf)
```

FIGURE 3.4: sf object with points.

3.5   Counting the number of points within polygons

We can use the st_intersects() function of **sf** to count the number of points within the polygons of a **sf** object. The returned object is a list with feature ids intersected in each of the polygons. We can use the lengths() function to calculate the number of points inside each feature as seen in Figure 3.5.

```
library(sf)
library(ggplot2)

# Map with divisions (sf object)
```

```
map <- read_sf(system.file("shape/nc.shp", package = "sf"))

# Points over map (simple feature geometry list-column sfc)
points <- st_sample(map, size = 100)

# Map of points within polygons
ggplot() + geom_sf(data = map) + geom_sf(data = points)

# Intersection (first argument map, then points)
inter <- st_intersects(map, points)

# Add point count to each polygon
map$count <- lengths(inter)

# Map of number of points within polygons
ggplot(map) + geom_sf(aes(fill = count))
```

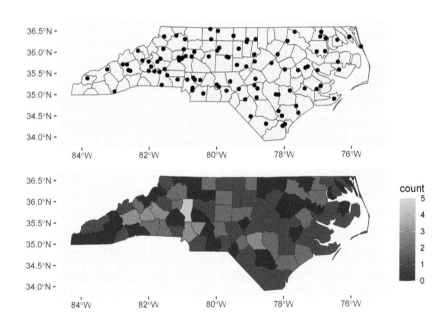

FIGURE 3.5: Top: Points within polygons. Bottom: Number of points within polygons.

3.6   Identifying polygons containing points

Given a `sf` object with points and a `sf` object with polygons, we can also use
the `st_intersects()` function to obtain the polygon each of the points belongs
to. For example, Figure 3.6 shows a map with the names of the polygons that
contain three points over the map.

```
library(sf)
library(ggplot2)

# Map with divisions (sf object)
map <- read_sf(system.file("shape/nc.shp", package = "sf"))

# Points over map (sf object)
points <- st_sample(map, size = 3) %>% st_as_sf()

# Intersection (first argument points, then map)
inter <- st_intersects(points, map)

# Adding column areaname with the name of
# the areas containing the points
points$areaname <- map[unlist(inter), "NAME",
                        drop = TRUE] # drop geometry
points
```

```
Simple feature collection with 3 features and 1 field
Geometry type: POINT
Dimension:     XY
Bounding box:  xmin: -80.18 ymin: 34.72 xmax: -77.91 ymax: 35.57
Geodetic CRS:  NAD27
                        x areaname
1 POINT (-80.18 35.57) Davidson
2 POINT (-79.26 34.72)  Robeson
3 POINT (-77.91 35.19)    Wayne
```

```
# Map
ggplot(map) + geom_sf() + geom_sf(data = points) +
  geom_sf_label(data = map[unlist(inter), ],
                aes(label = NAME), nudge_y = 0.2)
```

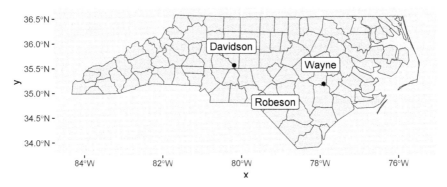

FIGURE 3.6: Map showing the name of polygons that contain three points over the map.

3.7 Joining map and data

Sometimes, a map and its corresponding data are available separately and we may wish to create a `sf` object representing the map with the added data that we can manipulate and plot. We can create a `sf` map with the data attributes by joining the map and the data with the `left_join()` function of the **dplyr** (Wickham et al., 2022b) package. Here, we present an example where we add air pollution data obtained with the **wbstats** (Piburn, 2020) package to a `sf` map obtained with the **rnaturalearth** package. First, we use the `ne_countries()` function of **rnaturalearth** (South, 2017) to download the world map with the country polygons of class `sf`.

```
library(rnaturalearth)
map <- ne_countries(returnclass = "sf")
```

Then, we use the **wbstats** package to download a data frame of air pollution data from the World Bank. Specifically, we search the pollution indicators with `wb_search()`, and use `wb_data()` to download PM2.5 in year 2016 by specifying the indicator corresponding to PM2.5, and the start and end dates.

```
library(wbstats)
indicators <- wb_search(pattern = "pollution")
d <- wb_data(indicator = "EN.ATM.PM25.MC.M3",
             start_date = 2016, end_date = 2016)
```

Then, we use the `left_join()` function of **dplyr** to join the map and the data specifying the argument `by` the variables we wish to join by. Here, we use ISO3 standard code of the countries rather than the country names, since names can be written differently in the map and the data frame. Figure 3.7 shows the map of the data obtain with **ggplot2**.

```
library(dplyr)
library(ggplot2)
library(viridis)

map1 <- left_join(map, d, by = c("iso_a3" = "iso3c"))
ggplot(map1) + geom_sf(aes(fill = EN.ATM.PM25.MC.M3)) +
  scale_fill_viridis() + labs(fill = "PM2.5") + theme_bw()
```

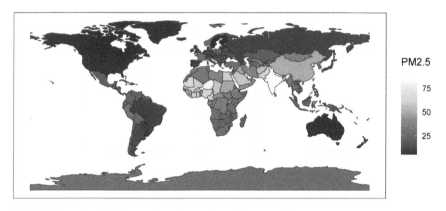

FIGURE 3.7: PM2.5 values in each of the world countries in 2016.

Note that when we use `left_join()`, the class of the result-ing object is the same as the class of the first argument. Thus, `left_join(sf_object, data.frame_object)` creates a `sf` object, whereas `left_join(data.frame_object, sf_object)` is a `data.frame`:

```
map1 <- left_join(map, d, by = c("iso_a3" = "iso3c"))
class(map1)
```

```
[1] "sf"            "data.frame"
```

```
d1 <- left_join(d, map, by = c("iso3c" = "iso_a3"))
class(d1)
```

```
[1] "tbl_df"      "tbl"            "data.frame"
```

Note also that given two data frames x and y, we can join them by using the `left_join()`, `right_join()`, `inner_join()` or `full_join()` functions.

Specifically, `left_join(x, y)` includes all rows in x so the resulting data frame includes all observations in the left data frame x, whether or not there is a match in the right data frame y. `right_join(x, y)` includes all rows in y so the resulting data frame includes all observations in y, whether or not there is a match in x. `inner_join(x, y)` includes all rows in x and y so the resulting data frame includes only observations that are in both data frames. Finally, `full_join(x, y)` includes all rows in x or y so the resulting data frame includes all observations from both data frames.

4

The **terra** package for raster and vector data

The **terra** package (Hijmans, 2022) has functions to create, read, manipulate, and write raster and vector data. Raster data is commonly used to represent spatially continuous phenomena by dividing the study region into a grid of equally sized rectangles (referred to as cells or pixels) with the values of the variable of interest. In **terra**, multilayer raster data is represented with the `SpatRaster` class. The `SpatVector` class is used to represent vector data such as points, lines and polygons, and their attributes. In this chapter, we provide examples on how to read, create, and operate with raster and vector data using **terra**.

4.1   Raster data

The `rast()` function can be used to create and read raster data. The `writeRaster()` function allows us to write raster data. For example, here we use `rast()` to read raster data representing elevation in Luxembourg from a file from the **terra** package, and assign it to an object `r` of class `SpatRaster` (Figure 4.1).

```
library(terra)
pathraster <- system.file("ex/elev.tif", package = "terra")
r <- rast(pathraster)
plot(r)
```

We can also use `rast()` to create a `SpatRaster` object `r` by specifying the number of columns, the number of rows, as well as the minimum and maximum x and y values.

```
r <- rast(ncol = 10, nrow = 10,
          xmin = -150, xmax = -80, ymin = 20, ymax = 60)
r
```

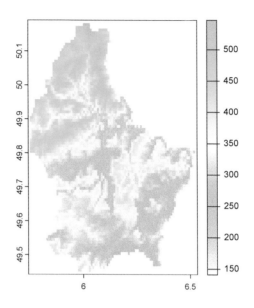

FIGURE 4.1: Elevation raster in Luxembourg obtained from **terra**.

```
class        : SpatRaster
dimensions   : 10, 10, 1  (nrow, ncol, nlyr)
resolution   : 7, 4  (x, y)
extent       : -150, -80, 20, 60  (xmin, xmax, ymin, ymax)
coord. ref.  : lon/lat WGS 84
```

We can get the size of the raster with several functions.

```
nrow(r) # number of rows
ncol(r) # number of columns
dim(r) # dimension
ncell(r) # number of cells
```

We use **values()** to set and access values of the raster.

```
values(r) <- 1:ncell(r)
```

We can also create a multilayer object using c().

```
r2 <- r * r
s <- c(r, r2)
```

Layers can be subsetted with `[[ ]]`.

```
plot(s[[2]]) # layer 2
```

Many generic functions can be used to operate with rasters as shown in the following examples.

```
plot(min(s))
plot(r + r + 10)
plot(round(r))
plot(r == 1)
```

4.2   Vector data

The class `SpatVector` of **terra** allows us to represent vector data such as points, lines, and polygons, as well as the attributes that describe these geometries. We can use `vect()` to read a shapefile, and `writeVector()` to write a `SpatVector` to a file. Here, we obtain the map with the divisions of Luxembourg that is in the shapefile file `lux.shp` of **terra**.

```
pathshp <- system.file("ex/lux.shp", package = "terra")
v <- vect(pathshp)
```

We can also use the `vect()` function to create a `SpatVector`. For example, here we create a `SpatVector` that contains longitude and latitude coordinates of point locations, and attributes representing character names and numeric values for each of the points.

```
# Longitude and latitude values
long <- c(-0.118092, 2.349014, -3.703339, 12.496366)
lat <- c(51.509865, 48.864716, 40.416729, 41.902782)
longlat <- cbind(long, lat)

# CRS
crspoints <- "+proj=longlat +datum=WGS84"

# Attributes for points
d <- data.frame(
place = c("London", "Paris", "Madrid", "Rome"),
```

```
value = c(200, 300, 400, 500))

# SpatVector object
pts <- vect(longlat, atts = d, crs = crspoints)

pts
plot(pts)
```

4.3   Cropping, masking, and aggregating raster data

The **terra** package provides several functions to work with raster data. Here, we show examples on how to crop, mask, and aggregate raster data by using a raster file representing temperature data. First, we use the `worldclim_country()` function of the **geodata** package (Hijmans et al., 2023) to download global temperature data from the WorldClim[1] database. Specifically, we download monthly average temperature in degree Celsius by specifying the country (`country = "Spain"`), the variable mean temperature (`var = "tavg"`), the resolution (`res = 10`), and the path where to download the data to as a temporary file (`path = tempdir()`). Figure 4.2 shows maps of the monthly average temperature in Spain.

```
library(terra)
r <- geodata::worldclim_country(country = "Spain", var = "tavg",
                                res = 10, path = tempdir())
plot(r)
```

We can average the temperature raster data over months with the `mean()` function (Figure 4.3).

```
r <- mean(r)
plot(r)
```

We also download the map of Spain with the **rnaturalearth** package (South, 2017), and delete the Canary Islands region (Figure 4.4).

```
# Map
library(ggplot2)
```

---

[1]https://www.worldclim.org/data/index.html

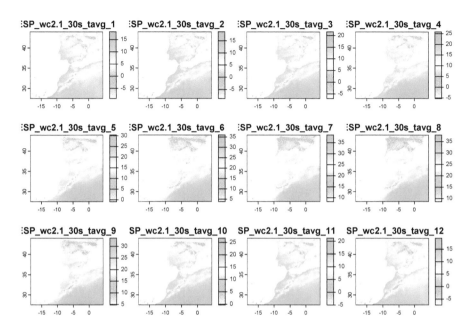

FIGURE 4.2: Monthly average temperature in Spain obtained from **geodata**.

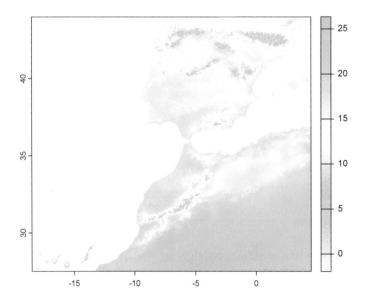

FIGURE 4.3: Raster representing the average annual temperature in Spain.

```
map <- rnaturalearth::ne_states("Spain", returnclass = "sf")
map <- map[-which(map$region == "Canary Is."), ] # delete region
ggplot(map) + geom_sf()
```

We obtain the spatial extent of the map with `terra:ext()`. Then, we can use `crop()` to remove the part of the raster that is outside the spatial extent (Figure 4.4).

```
# Cropping
sextent <- terra::ext(map)
r <- terra::crop(r, sextent)
plot(r)
```

We can use the `mask()` function to convert all values outside the map to NA (Figure 4.4).

```
# Masking
r <- terra::mask(r, vect(map))
plot(r)
```

The `aggregate()` function of **terra** can be used to aggregate groups of cells of a raster in order to create a new raster with a lower resolution (i.e., larger cells). The argument `fact` of `aggregate()` denotes the aggregation factor expressed as number of cells in each direction (horizontally and vertically), or two integers denoting the horizontal and vertical aggregation factor. Argument `fun` specifies the function used to aggregate values (e.g., `mean`). Figure 4.4 shows a low resolution raster of average annual temperatures in Spain obtained using the `aggregate()` function.

```
# Aggregating
r <- terra::aggregate(r, fact = 20, fun = "mean", na.rm = TRUE)
plot(r)
```

4.4   Extracting raster values at points

Given a raster of class `SpatRaster`, we can extract the raster values at a set of points with the `extract()` function of **terra**. Here, we provide an example of the use of `extract()` using a raster representing the elevation of Luxembourg,

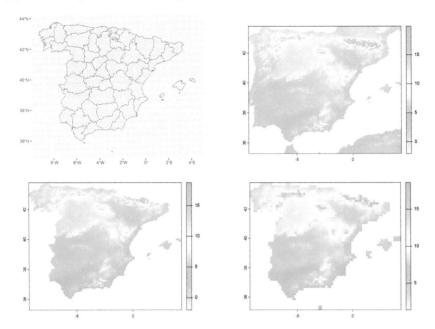

FIGURE 4.4: Map of Spain excluding the Canary Islands region (top-left), and cropped raster (top-right), masked raster (bottom-left), and low resolution raster (bottom-right) representing the average annual temperature in the map.

and a vector file with the divisions of Luxembourg from files in the **terra** package.

```
library(terra)
# Raster (SpatRaster)
r <- rast(system.file("ex/elev.tif", package = "terra"))
# Polygons (SpatVector)
v <- vect(system.file("ex/lux.shp", package = "terra"))
```

We use the **terra** functions `centroids()` to obtain the centroids of the division polygons, and `crds()` to obtain their coordinates.

```
points <- crds(centroids(v))
```

Figure 4.5 shows the elevation raster, the polygons, and the points of the centroid locations.

```
plot(r)
plot(v, add = TRUE)
points(points)
```

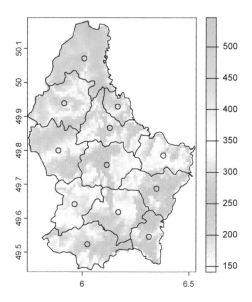

FIGURE 4.5: Elevation raster, and division and centroids of polygons in Luxembourg.

Now, we obtain the values of the raster at `points` using `extract()` passing as first argument the `SpatVector` object with the raster, and as second argument a data frame with the points.

```
# data frame with the coordinates
points <- as.data.frame(points)
valuesatpoints <- extract(r, points)
cbind(points, valuesatpoints)
```

```
        x      y ID elevation
1   6.009 50.07  1       444
2   6.127 49.87  2       295
3   5.887 49.80  3       382
4   6.165 49.93  4       404
5   5.915 49.94  5       414
6   6.378 49.79  6       320
```

```
7   6.312 49.55  7         193
8   6.346 49.69  8         228
9   5.964 49.64  9         313
10 6.024 49.52 10         282
11 6.168 49.62 11         328
12 6.114 49.76 12         221
```

4.5 Extracting and averaging raster values within polygons

We can also use `extract()` to obtain the values of raster objects of class `SpatRaster` within polygons of class `SpatVector`. By default, cells extracted within each polygon are cells that have centers covered by the polygon. We can set the argument `weights = TRUE` to get, apart from the cell values, the percentage of each cell covered by the polygon, and this can be used to compute area-weighted averages. The argument `fun` of `extract()` can be used to specify a function (e.g., `mean`) that summarizes the extracted values by polygon.

```
# Extracted raster cells within each polygon
head(extract(r, v, na.rm = TRUE))
```

```
  ID elevation
1  1       547
2  1       485
3  1       497
4  1       515
5  1       515
6  1       515
```

```
# Extracted raster cells and percentage of area
# covered within each polygon
head(extract(r, v, na.rm = TRUE, weights = TRUE))
```

```
  ID elevation  weight
1  1        NA 0.04545
2  1        NA 0.10909
3  1       529 0.24545
4  1       542 0.46364
5  1       547 0.68182
6  1       535 0.11818
```

Thus, the average raster values by polygon are obtained with `extract()` as follows.

```
# Average raster values by polygon
v$avg <- extract(r, v, mean, na.rm = TRUE)$elevation
```

The area-weighted average raster values by polygon are obtained with `extract()` setting `weights = TRUE`.

```
# Area-weighted average raster values by polygon (weights = TRUE)
v$weightedavg <- extract(r, v, mean, na.rm = TRUE,
                 weights = TRUE)$elevation
```

Figure 4.6 shows maps of the elevation averages and area-weighted averages in each of the Luxembourg divisions created with the **ggplot2** (Wickham et al., 2022a) and **tidyterra** (Hernangomez, 2023b) packages.

```
library(ggplot2)
library(tidyterra)

# Plot average raster values within polygons
ggplot(data = v) + geom_spatvector(aes(fill = avg)) +
  scale_fill_terrain_c()

# Plot area-weighted average raster values within polygons
ggplot(data = v) + geom_spatvector(aes(fill = weightedavg)) +
  scale_fill_terrain_c()
```

FIGURE 4.6: Average and area-weighted average of elevation values in each of the divisions of Luxembourg.

5

Making maps with R

Maps allow us to easily convey spatial information. Here, we show how to create both static and interactive maps by using several mapping packages including **ggplot2** (Wickham et al., 2022a), **leaflet** (Cheng et al., 2022a), **mapview** (Appelhans et al., 2022), and **tmap** (Tennekes, 2022). We create maps of areal data using several functions and parameters of the mapping packages. We also briefly describe how to plot point and raster data. Then, we show how to create maps of flows between locations with the **flowmapblue** package (Boyandin, 2023).

The areal data we map correspond to sudden infant deaths in the counties of North Carolina, USA, in 1974 and 1979 which are in the **sf** package (Pebesma, 2022a). The path of the data can be obtained with `system.file()` specifying the directory of the data (`"shape/nc.shp"`) and the package (`"sf"`). Then, the data can be read with the `st_read()` function of **sf** by specifying the path of the data. We call this data `d`, and create variables `vble` and `vble2` that we wish to map with the values of the sudden infant deaths in 1974 and 1979, respectively.

```
library(sf)
nameshp <- system.file("shape/nc.shp", package = "sf")
d <- st_read(nameshp, quiet = TRUE)
d$vble <- d$SID74
d$vble2 <- d$SID79
```

5.1 ggplot2

The **ggplot2** package (Wickham et al., 2022a) allows us to create graphics based on the grammar of graphics that defines the rules of structuring mathematic and aesthetic elements to build graphs layer-by-layer.

To create a plot with **ggplot2**, we call `ggplot()` specifying arguments `data` which is a data frame with the variables to plot, and `mapping = aes()` which are aesthetic mappings between variables in the data and visual properties of the objects in the graph such as position and color of points or lines.

Then, we use + to add layers of graphical components to the graph. Layers consist of geometries, stats, scales, coordinates, facets, and themes. For example, we add objects to the graph with geom_*() functions (e.g, geom_point() for points, geom_line() for lines). We can also add color scales (e.g., scale_colour_brewer()), faceting specifications (e.g., facet_wrap() splits data into subsets to create multiple plots), and coordinate systems (e.g., coord_flip()).

We can create maps by using the geom_sf() function and providing a simple feature (sf) object. Figure 5.1 shows the map of sudden infant deaths in North Carolina in 1974 (vble) created with **ggplot2** with viridis scale from the **viridis** package (Garnier, 2021).

```
library(ggplot2)
library(viridis)
ggplot(d) + geom_sf(aes(fill = vble)) +
  scale_fill_viridis() + theme_bw()
```

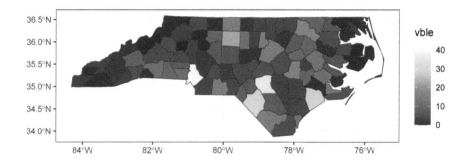

FIGURE 5.1: Map of areal data created with **ggplot2**.

Plots created with **ggplot2** can be saved with the ggsave() function. Alternatively, we can specify a device driver (e.g., png, pdf), print the plot, and then shut down the device with dev.off():

```
png("plot.png")
ggplot(d) + geom_sf(aes(fill = vble)) +
  scale_fill_viridis() + theme_bw()
dev.off()
```

Moreover, the **plotly** package (Sievert et al., 2022b) can be used in combination with **ggplot2** to create an interactive plot. Specifically, we can turn a static ggplot object to an interactive `plotly` object by calling the `ggplotly()` function of **plotly** providing the `ggplot` object (Figure 5.2).

```
library(plotly)
g <- ggplot(d) + geom_sf(aes(fill = vble))
ggplotly(g)
```

FIGURE 5.2: Interactive map of areal data created with **plotly**.

The **gganimate** package (Pedersen and Robinson, 2022) can also be used to create animated plots. The syntax of this package is similar to that of **ggplot2** and has additional functions to define how data should change in the animation.

5.2 leaflet

The **leaflet** package (Cheng et al., 2022a) makes it easy to create maps using Leaflet[1] which is a very popular open-source JavaScript library for interactive

---

[1]https://leafletjs.com/

maps. We can create a leaflet map by calling the `leaflet()` function passing the spatial object, and adding layers such as polygons and legends using a number of functions. The **sf** object that we pass to `leaflet()` needs to have a geographic coordinate reference system (CRS) indicating latitude and longitude (EPSG code 4326). Here, we use the `st_transform()` function of **sf** to transform the data **d** which has CRS given by EPSG code 4267 to CRS with EPSG code 4326.

```
st_crs(d)$epsg
```

```
[1] 4267
```

```
d <- st_transform(d, 4326)
```

We create a color palette using the `colorNumeric()` function of **leaflet** specifying the color palette `"YlOrRd"` of the **RColorBrewer** package (Neuwirth, 2022), and the domain with the possible values that can be mapped (**d$vble**).

Then, we create the map using `leaflet()` and `addTiles()` to add a background map to put data into context. Then, we use `addPolygons()` to add the polygons representing counties specifying the color of the border (`color`), interior (`fillColor`) and opacity (`fillOpacity`). Finally, we add a legend with `addLegend()` (Figure 5.3).

```
library(leaflet)
pal <- colorNumeric(palette = "YlOrRd", domain = d$vble)
l <- leaflet(d) %>% addTiles() %>%
  addPolygons(color = "white", fillColor = ~ pal(vble),
              fillOpacity = 0.8) %>%
  addLegend(pal = pal, values = ~vble, opacity = 0.8)
l
```

Note that the default background map added with `addTiles()` can be changed by another map with `addProviderTiles()` specifying another tile layer. Examples of tile layers can be seen at the leaflet providers' website[2]. We can also use the `addMiniMap()` function to add an inset map (Figure 5.4).

```
l %>% addMiniMap()
```

The `saveWidget()` function of **htmlwidgets** (Vaidyanathan et al., 2023) allows us to save the map created to an HTML file. If we wish to save an image file, we can first save the leaflet map as an HTML file with `saveWidget()`,

---

[2]http://leaflet-extras.github.io/leaflet-providers/preview/index.html

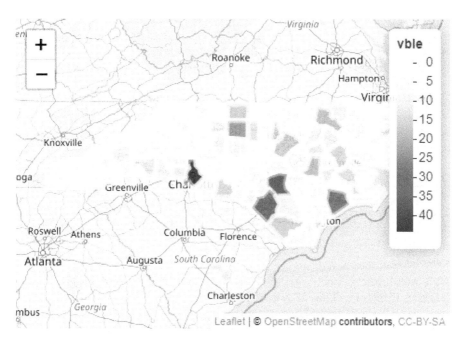

FIGURE 5.3: Map of areal data created with **leaflet**.

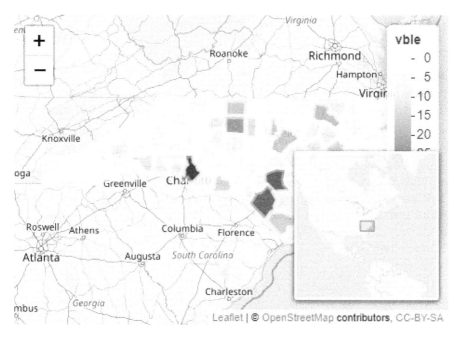

FIGURE 5.4: Map of areal data with inset map created with **leaflet**.

and then capture a static version of the HTML using the `webshot()` function of the **webshot** package (Chang, 2022a). The use of **webshot** requires the installation of the external program `PhantomJS` which can be installed with `webshot::install_phantomjs()`. For example, to save the leaflet map 1 as `.png` we can proceed as follows:

```
# Saves map.html
library(htmlwidgets)
saveWidget(widget = 1, file = "map.html")

# Takes a screenshot of the map.html created
# and saves it as map.png
library(webshot)
# webshot::install_phantomjs()
webshot(url = "map.html", file = "map.png")
```

Note we can use `getwd()` and `setwd()` to get and set, respectively, the current working directory. This allows us to see where the files were saved. Moreover, if in `saveWidget()` we specify the path where to save `map.html` as `file = "directory/map.html"`, the same path needs to be specified in the argument `url` of `webshot()` as `url = "directory/map.html"`. The package **webshot2** (Chang, 2022b) is meant to be a replacement for **webshot**. The **webshot2** package uses headless Chrome via the **chromote** package (Chang and Schloerke, 2022), whereas **webshot** uses `PhantomJS`.

5.3   mapview

The **mapview** package (Appelhans et al., 2022) allows us to very quickly create similar interactive maps as **leaflet**. Specifically, we just need to use the `mapview()` function passing as arguments the spatial object and the variable we want to plot (`zcol`). Figure 5.5 shows the map created with **mapview**. This map is interactive and by clicking each of the areas we can see popups with the data information.

```
library(mapview)
mapview(d, zcol = "vble")
```

Maps created with **mapview** can also be customized to add elements such as legends and background maps. For example, we can choose another background map by using the argument `map.types`, change the color palette with `col.regions`, and put a title for the legend with `layer.name` as follows:

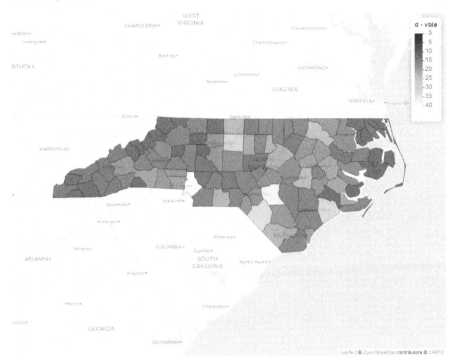

FIGURE 5.5: Map of areal data created with **mapview**.

```
library(RColorBrewer)
pal <- colorRampPalette(brewer.pal(9, "YlOrRd"))
mapview(d, zcol = "vble", map.types = "CartoDB.DarkMatter",
        col.regions = pal, layer.name = "SDI")
```

An inset map can also be added by using the addMiniMap() function of **leaflet**.

```
map1 <- mapview(d, zcol = "vble")
leaflet::addMiniMap(map1@map)
```

We can save maps created with **mapview** by using the mapshot() function of **mapview** as an HTML file or as a PNG, PDF, or JPEG image.

5.4   Side-by-side plots with mapview

We can create a side-by-side plot of maps obtained with **mapview** separated
with a slider by using the | operator. To use the | operator, we need to install
the **leaflet.extras2** package (Sebastian, 2023). When creating the individual
maps with mapview(), we use a common legend by specifying the same color
palette (col.regions) and breaks (at) so in both maps colors correspond to
the same intervals of values (Figure 5.6).

```
library(leaflet.extras2)
library(RColorBrewer)
pal <- colorRampPalette(brewer.pal(9, "YlOrRd"))

# common legend
at <- seq(min(c(d$vble, d$vble2)), max(c(d$vble, d$vble2)),
          length.out = 8)

m1 <- mapview(d, zcol = "vble", map.types = "CartoDB.Positron",
              col.regions = pal, at = at)
m2 <- mapview(d, zcol = "vble2", map.types = "CartoDB.Positron",
              col.regions = pal, at = at)

m1 | m2
```

5.5   Synchronized maps with leafsync

The **leafsync** package (Appelhans and Russell, 2019) can be used to produce
a lattice-like view of multiple synchronized maps created with **mapview** or
leaflet. Here, we show how to create maps of sudden infant deaths in 1974
(vble) and 1979 (vble2) with synchronized zoom and pan. First, we use
mapview() to create maps of the variables vble and vble2. Then, we use the
sync() function of **leafsync** passing the maps created (Figure 5.7).

```
library(RColorBrewer)
pal <- colorRampPalette(brewer.pal(9, "YlOrRd"))

# common legend
at <- seq(min(c(d$vble, d$vble2)), max(c(d$vble, d$vble2)),
```

FIGURE 5.6: Side-by-side maps created with **mapview** and the | operator of **leaflet.extras2**.

```
          length.out = 8)

m1 <- mapview(d, zcol = "vble", map.types = "CartoDB.Positron",
              col.regions = pal, at = at)
m2 <- mapview(d, zcol = "vble2", map.types = "CartoDB.Positron",
              col.regions = pal, at = at)


m <- leafsync::sync(m1, m2)
m
```

This synchronized map can be saved by using the `save_html()` function of the **htmltools** package (Cheng et al., 2022b).

```
htmltools::save_html(m, "m.html")
```

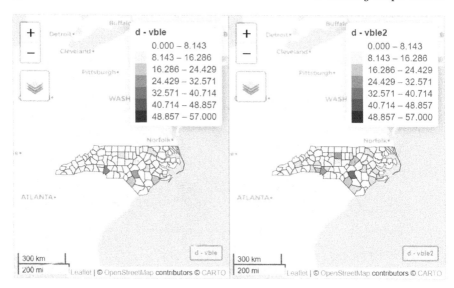

FIGURE 5.7: Synchronized maps created with **leafsync**.

5.6   tmap

The **tmap** package (Tennekes, 2022) allows us to create static and interactive maps composed of multiple shapes and layers, and with different styles. Maps are created with `tm_shape()` specifying the `sf` object. Then, layers are added with a `tm_*()` function. For example, `tm_polygons()` draws polygons, `tm_dots()` draws dots, and `tm_text()` adds text. Functions `tmap_mode("plot")` and `tmap_mode("view")` can be used to set the static and interactive modes of the maps, respectively. Figure 5.8 shows a static map created with **tmap** with the values corresponding to areal data.

```
library(tmap)
tmap_mode("plot")
tm_shape(d) + tm_polygons("vble")
```

The function `tmap_save()` can be used to save maps by specifying the name of the HTML file (if `view` mode) or image (if `plot` mode).

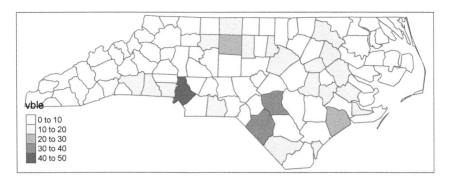

FIGURE 5.8: Map of areal data created with **tmap**.

5.7 Maps of point data

In addition to mapping areal data, the **ggplot2**, **leaflet**, **mapview** and **tmap** packages can also be used to create maps of point and raster data. Here, we show how to use these packages to represent the locations and population sizes of South African cities using points with different colors and sizes. These data are obtained from the `world.cities` data from the **maps** package (Brownrigg, 2022).

```
library(maps)
d <- world.cities
# Select South Africa
d <- d[which(d$country.etc == "South Africa"), ]
# Transform data to sf object
d <- st_as_sf(d, coords = c("long", "lat"))
# Assign CRS
st_crs(d) <- 4326
```

We create the variables `vble` with the population, and `size` with a transformation of the population that will be used to plot the points.

```
d$vble <- d$pop
d$size <- sqrt(d$vble)/100
```

Figure 5.9 shows a map created with **ggplot2** by setting the points color with `col = vble` and their size with `size = size`.

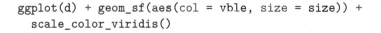

```
ggplot(d) + geom_sf(aes(col = vble, size = size)) +
  scale_color_viridis()
```

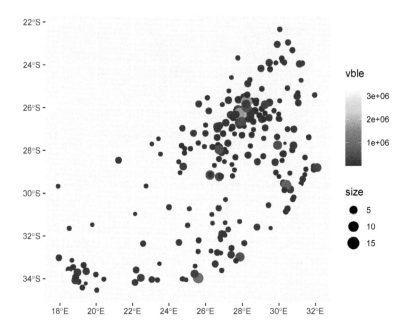

FIGURE 5.9: Map of point data created with **ggplot2**.

A leaflet map can be created using `addCircles()` specifying the radius and color of the points (Figure 5.10).

```
pal <- colorNumeric(palette = "viridis", domain = d$vble)
leaflet(d) %>% addTiles() %>%
  addCircles(lng = st_coordinates(d)[, 1],
             lat = st_coordinates(d)[, 2],
             radius = ~sqrt(vble)*10,
             color = ~pal(vble), popup = ~name) %>%
  addLegend(pal = pal, values = ~vble, position = "bottomright")
```

To create a map with **mapview**, we set the size of the points with `cex = "size"`.

```
d$size <- sqrt(d$vble)
mapview(d, zcol = "vble", cex = "size")
```

Finally, we use `tm_dots()` to create a map with **tmap**.

```
tmap_mode("view")
tm_shape(d) + tm_dots("vble", scale = sqrt(d$vble)/500,
                      palette = "viridis")
```

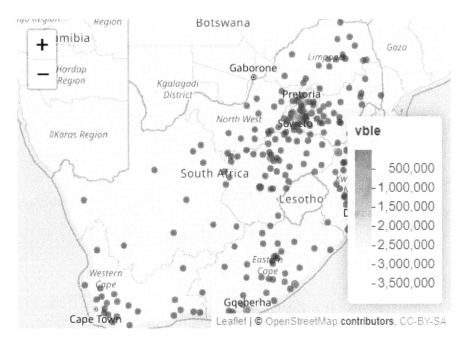

FIGURE 5.10: Map of point data created with **leaflet**.

5.8 Maps of raster data

These packages can also be used to plot raster data. Here, we plot the raster data representing elevation in Luxembourg obtained from the **terra** package (Hijmans, 2022).

```
library(terra)
filename <- system.file("ex/elev.tif", package = "terra")
r <- rast(filename)
```

To create a map using **ggplot2**, we transform the data to a data frame and pass it to `geom_raster()` (Figure 5.11).

```
# Transform data to sf object
d <- st_as_sf(as.data.frame(r, xy = TRUE), coords = c("x", "y"))
# Assign CRS
st_crs(d) <- 4326
# Plot
ggplot(d) + geom_sf() +
  geom_raster(data = as.data.frame(r, xy = TRUE),
              aes(x = x, y = y, fill = elevation))
```

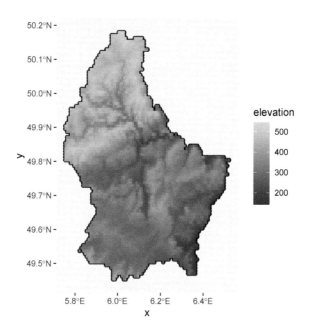

FIGURE 5.11: Map of raster data created with **ggplot2**.

To use the **leaflet** and **mapview** packages, we transform the data from class
`terra` to `RasterLayer` with the `raster::brick()` function. Figure 5.12 shows
the map of raster data created with **leaflet**.

```
library(raster)
rb <- raster::brick(r)

pal <- colorNumeric("YlOrRd", values(r),
                    na.color = "transparent")
leaflet() %>% addTiles() %>%
  addRasterImage(rb, colors = pal, opacity = 0.8) %>%
  addLegend(pal = pal, values = values(r), title = "elevation")
```

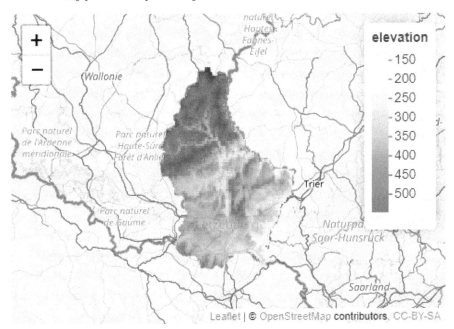

FIGURE 5.12: Map of raster data created with **leaflet**.

We create a map with **mapview** setting the title with the argument `layer` as follows:

```
mapview(rb, layer = "elevation")
```

We create a map with **tmap** by using `tm_raster()`.

```
tm_shape(r) + tm_raster(title = "elevation", palette = "YlOrRd")
```

5.9 Mobility flows with flowmapblue

The **flowmapblue** package (Boyandin, 2023) is a FlowmapBlue[3] widget for R that can be used to easily map mobility data between locations. An example on the use of **flowmapblue** showing population flows in Spain derived from

---

[3]http://flowmap.blue

cellphone location data can be seen at this blog post[4]. Flow data can also be depicted with the visualization functions of the **epiflows** package for the prediction of the spread of infectious diseases (Piatkowski et al., 2018; Moraga et al., 2019).

To use **flowmapblue**, we need to create a token at https://account.mapbox.com/. We can create an interactive mobility map by using just a few lines of code. First, we need to install the package from GitHub as follows:

```
devtools::install_github("FlowmapBlue/flowmapblue.R")
library(flowmapblue)
```

Then, we need to create two data frames containing the locations and the flows between locations. The data frame `locations` contains the ids, names, and coordinates of each of the locations. For example:

```
locations <- data.frame(
id = c(1, 2, 3),
name = c("New York", "London", "Rio de Janeiro"),
lat = c(40.713543, 51.507425, -22.906241),
lon = c(-74.011219, -0.127738, -43.180244)
)
```

The data frame `flows` has the number of people moving between origin and destination locations. For example:

```
flows <- data.frame(
origin = c(1, 2, 3, 2, 1, 3),
dest = c(2, 1, 1, 3, 3 , 2),
count = c(42, 51, 50, 40, 22, 42)
)
```

Finally, we call the `flowmapblue()` function passing the data frames `locations` and `flows`, the mapbox access token and specifying several options such as clustering or animation. Note that by setting `mapboxAccessToken = NULL`, we will obtain a visualization of the flows between locations but without a background map.

```
flowmapblue(locations, flows, mapboxAccessToken,
            clustering = TRUE, darkMode = TRUE, animation = FALSE)
```

---

[4]https://www.paulamoraga.com/blog/2020/07/11/2020-07-11-mobility/

Figure 5.13 shows a screenshot of an interactive mobility map created with **flowmapblue**. This map can be explored by moving the map, zooming in and out, and clicking the arrows to see the movement associated to each flow. We can also click the bottom right corner to open the map in full screen mode.

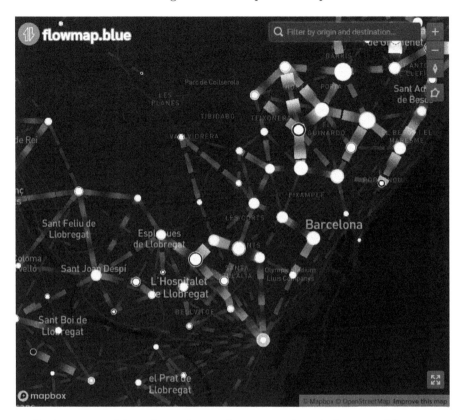

FIGURE 5.13: Map of population flows created with **flowmapblue**.

6

R packages to download open spatial data

Spatial data are used in a wide range of disciplines including environment, health, agriculture, economy, and society (Moraga and Baker, 2022). Several R packages have been recently developed as clients for various databases that can be used for easy access of spatial data including administrative boundaries, climatic, and OpenStreetMap data. Here, we give short reproducible examples on how to download and visualize spatial data that can be useful in different settings. More extended examples and details about the capabilities of each of the packages can be seen at the packages' websites, and the rspatialdata[1] website which provides a collection of tutorials on R packages to download and visualize spatial data using R.

6.1 Administrative boundaries of countries

We can download administrative boundaries of world countries with the **rnaturalearth** package (South, 2017). Other packages can also be used to obtain data of specific countries such as the USA with **tidycensus** (Walker and Herman, 2023) and **tigris** (Walker, 2023), Spain with **mapSpain** (Hernangomez, 2022), and Brazil with **geobr** (Pereira and Goncalves, 2022). The **giscoR** package (Hernangomez, 2023a) helps to retrieve data from Eurostat - GISCO (the Geographic Information System of the COmmission)[2] which contains several open data such as countries and coastal lines.

Here, we use **rnaturalearth** to download the administrative boundaries from Natural Earth map data[3]. Note that when installing **rnaturalearth**, we may get an error that can be fixed by installing the **rnaturalearthhires** package (South, 2023). The `ne_countries()` function allows us to download the map of the country specified in argument `country`, of scale given in `scale`, and of class `sp` or `sf` given in `returnclass`. We can retrieve the possible names that can be specified in argument `country` by typing `ne_countries()$admin`.

---

[1] https://rspatialdata.github.io/
[2] https://ec.europa.eu/eurostat/web/gisco
[3] https://www.naturalearthdata.com/

The `ne_states()` function can be used to obtain administrative divisions for specific countries.

Figure 6.1 shows maps of Germany and its divisions downloaded with **rnaturalearth**, and plotted side-by-side with the **patchwork** package (Pedersen, 2022).

```
# install.packages("devtools")
# devtools::install_github("ropensci/rnaturalearthhires")

library(rnaturalearth)
library(sf)
library(ggplot2)
library(viridis)
library(patchwork)

map1 <- ne_countries(type = "countries", country = "Germany",
                     scale = "medium", returnclass = "sf")
map2 <- rnaturalearth::ne_states("Germany", returnclass = "sf")
p1 <- ggplot(map1) + geom_sf()
p2 <- ggplot(map2) + geom_sf()
p1 + p2
```

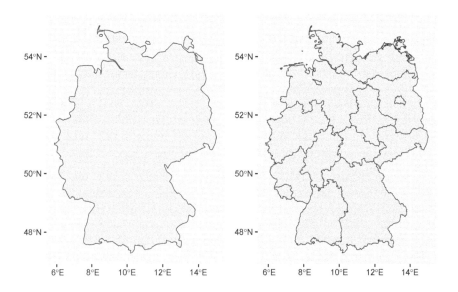

FIGURE 6.1: Maps of Germany and its divisions obtained with **rnaturalearth**.

Note that in case of needing maps of administrative boundaries at greater levels than the ones provided by these packages, we could resort to other sources such as online data repositories maintained by specific countries.

6.2 Climatic data

The **geodata** package (Hijmans et al., 2023) allows us to download geographic data including climate, elevation, land use, soil, crop, species occurrence, administrative boundaries, and other data. The **geodata** package is a successor of the `getData()` function from the **raster** package (Hijmans, 2023).

For example, the `worldclim_country()` function downloads climate data from WorldClim[4] including minimum temperature (`tmin`), maximum temperature (`tmax`), average temperature (`tavg`), precipitation (`prec`), and wind speed (`wind`). The `country_codes()` function of **geodata** can be used to get the names and codes of the world countries. Here, we provide an example on how to download minimum temperature in Jamaica using `worldclim_country()` specifying `country = "Jamaica"`, `var = "tmin"` and `path = tempdir()` as the path name of the temporary directory to download the data. This function retrieves the temperature for each month, and we can plot the mean over the months with `mean(d)` (Figure 6.2).

```
library(geodata)
d <- worldclim_country(country = "Jamaica", var = "tmin",
                       path = tempdir())
terra::plot(mean(d), plg = list(title = "Min. temperature (C)"))
```

The **geodata** package can also be used to download other data such as elevation with `elevation_30s()`, land cover with `landcover()`, and soil with `soil_world()`.

6.3 Precipitation

The **chirps** package (de Sousa et al., 2022) allows us to obtain daily high-resolution precipitation, as well as daily maximum and minimum temperatures from the Climate Hazards Group[5]. We use the `get_chirps()` function to obtain daily precipitation in Bangkok, Thailand, by specifying the longitude

---

[4]https://www.worldclim.org/
[5]https://www.chc.ucsb.edu/

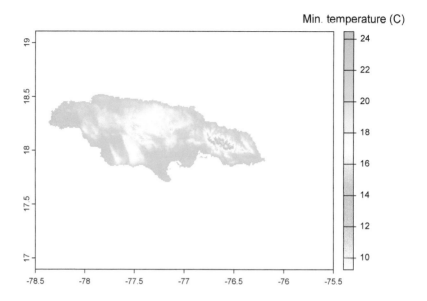

FIGURE 6.2: Minimum temperature in Jamaica obtained with **geodata**.

and latitude coordinates of Bangkok, the dates, and the server source. Here, we use the `"ClimateSERV"` server instead of the default server `"CHC"`, since it is recommended when few data points are required (Figure 6.3).

```
library("chirps")
location <- data.frame(long = 100.523186, lat = 13.736717)
d <- get_chirps(location, dates = c("2020-01-01", "2022-12-31"),
                server = "ClimateSERV")
ggplot(d, aes(x = date, y = chirps)) + geom_line() +
  labs(y = "Precipitation (mm)")
```

6.4   Elevation

The **elevatr** package (Hollister, 2022) allows us to get elevation data from Amazon Web Services (AWS) Terrain Tiles[6] and OpenTopography Global Digital Elevation Models API[7]. The `get_elev_raster()` function can be used to download elevation at the locations specified in argument `locations` and

---

[6]https://registry.opendata.aws/terrain-tiles/
[7]https://portal.opentopography.org/apidocs/#/Public/getGlobalDem

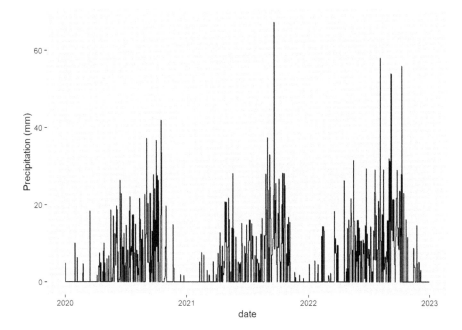

FIGURE 6.3: Daily precipitation in Bangkok obtained with **chirps**.

with a zoom specified in argument `z`. Argument `clip` can be set to `"tile"` to return full tiles, `"bbox"` to return data clipped to the bounding box of the locations, or `"locations"` to return data clipped to the data specified in `locations`. Figure 6.4 shows the elevation of Switzerland downloaded passing to `get_elev_raster()` a `sf` object with the map of the country.

```
library(rnaturalearth)
library(elevatr)
library(terra)
map <- ne_countries(type = "countries", country = "Switzerland",
                    scale = "medium", returnclass = "sf")
d <- get_elev_raster(locations = map, z = 9, clip = "locations")
terra::plot(rast(d), plg = list(title = "Elevation (m)"))
```

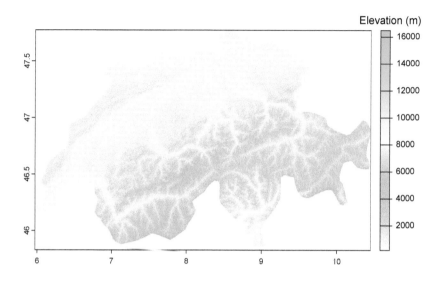

FIGURE 6.4: Elevation in Switzerland obtained with **elevatr**.

6.5   OpenStreetMap data

OpenStreetMap (OSM)[8] is an open world geographic database updated and maintained by a community of volunteers. We can use the **osmdata** package (Padgham et al., 2023) to retrieve OSM data including roads, shops, railway stations, and much more. The `available_features()` function can be used to get the list of recognized features in OSM. This list can be found in the OSM wiki[9].

```
library(osmdata)
head(available_features())
```

```
[1] "4wd_only"  "abandoned" "abutters"  "access"
[5] "addr"      "addr:city"
```

The `available_tags()` function lists out the tags associated with each feature. For example, tags associated with feature `"amenity"` can be obtained as follows:

---

[8]https://www.openstreetmap.org/
[9]https://wiki.openstreetmap.org/wiki/Map_Features

```
head(available_tags("amenity"))
```

```
# A tibble: 6 x 2
  Key     Value
  <chr>   <chr>
1 amenity [[ Data item not found. Check your spelling.~
2 amenity animal_boarding
3 amenity animal_breeding
4 amenity animal_shelter
5 amenity animal_training
6 amenity arts_centre
```

The first step in creating an `osmdata` query is defining the geographical area we wish to include in the query. This can be done by defining a bounding box that defines a geographical area by its bounding latitudes and longitudes. The bounding box for a given place name can be obtained with the `getbb()` function. For example, the bounding box of Barcelona, Spain, can be obtained as follows.

```
placebb <- getbb("Barcelona")
placebb
```

```
       min    max
x    2.052  2.228
y   41.317 41.468
```

To retrieve the required features of a place defined by the bounding box, we need to create an overpass query with `opq()`. Then, the `add_osm_feature()` function can be used to add the required features to the query. Finally, we use the `osmdata_sf()` function to obtain a simple feature object of the resultant query. For example, we can obtain the hospitals of Barcelona by specifying its bounding box `placebb` and using `add_osm_feature()` with `key = "amenity"` and `value = "hospital"` as follows.

```
hospitals <- placebb %>% opq() %>%
  add_osm_feature(key = "amenity", value = "hospital") %>%
  osmdata_sf()
```

Motorways can be retrieved using `key = "highway"` and `value = "motorway"`.

```
motorways <- placebb %>% opq() %>%
```

```
add_osm_feature(key = "highway", value = "motorway") %>%
osmdata_sf()
```

Figure 6.5 shows an interactive map with the hospitals and motorways of
Barcelona downloaded with **osmdata**.

```
library(leaflet)
leaflet() %>% addTiles() %>%
  addPolylines(data = motorways$osm_lines, color = "black") %>%
  addPolygons(data = hospitals$osm_polygons,
              label = hospitals$osm_polygons$name)
```

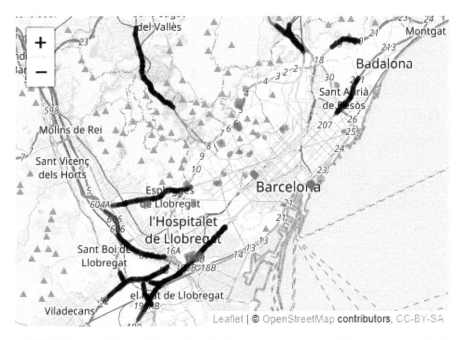

FIGURE 6.5: Map with the hospitals and motorways of Barcelona obtained
with **osmdata**.

6.6 World Bank data

The World Bank[10] provides a great source of global socio-economic data
spanning several decades and dozens of topics, with the potential to shed light

---

[10]https://www.worldbank.org/

on numerous global issues. Some of the indicators can be seen at this website[11]. The **wbstats** package (Piburn, 2020) allows us to search and download data from the World Bank API. The `wb_search()` function can be used to find indicators that match a search term. For example, we can find indicators that contain the words `"poverty"` or `"unemployment"` as follows.

```
library(wbstats)
indicators <- wb_search(pattern = "poverty|unemployment")
# print(indicators)
```

We can inspect the indicators retrieved with `View(indicators)`. The function `wb_data()` allows us to retrieve the chosen data. For example, here we download Human Development Index which has ID `MO.INDEX.HDEV.XQ` in 2011.

```
d <- wb_data(indicator = "MO.INDEX.HDEV.XQ",
             start_date = 2011, end_date = 2011)
print(head(d))
```

```
# A tibble: 6 x 9
  iso2c iso3c country        date MO.INDEX.HDEV.XQ unit
  <chr> <chr> <chr>         <dbl>            <dbl> <chr>
1 AO    AGO   Angola         2011             47.7 <NA>
2 BI    BDI   Burundi        2011             48.5 <NA>
3 BJ    BEN   Benin          2011             53.2 <NA>
4 BF    BFA   Burkina Faso   2011             45.8 <NA>
5 BW    BWA   Botswana       2011             80.3 <NA>
6 CF    CAF   Central Afr~   2011             32.9 <NA>
# i 3 more variables: obs_status <chr>,
#   footnote <chr>, last_updated <date>
```

We can visualize the data by adding the data d to a map of Africa retrieved with the `ne_countries()` function of **rnaturalearth** (Figure 6.6).

```
library(rnaturalearth)
library(mapview)
map <- ne_countries(continent = "Africa", returnclass = "sf")
map <- dplyr::left_join(map, d, by = c("iso_a3" = "iso3c"))
mapview(map, zcol = "MO.INDEX.HDEV.XQ")
```

---

[11]https://data.worldbank.org/indicator

FIGURE 6.6: Human Development Index obtained with **wbstats**.

6.7   Species occurrence

The **spocc** package (Chamberlain, 2021) is an interface to many species occurrence data sources including Global Biodiversity Information Facility (GBIF), USGSs' Biodiversity Information Serving Our Nation (BISON), iNaturalist, eBird, Integrated Digitized Biocollections (iDigBio), VertNet, Ocean Biogeographic Information System (OBIS), and Atlas of Living Australia (ALA). The package provides functionality to retrieve and combine species occurrence data.

The occ() function from **spocc** can be used to retrieve the locations of species. Here, we download data on brown-throated sloths in Costa Rica recorded between 2000 and 2019 from the GBIF database. Arguments of this function include query with the species scientific name (*Bradypus variegatus*), from with the name of the database (GBIF), and date with the start and end dates (2000-01-01 to 2019-12-31). We also specify that we wish to retrieve occurrences in Costa Rica by setting gbifopts to a named list with country equal to the 2-letter code of Costa Rica (CR). Moreover, we only retrieve occurrence data

that have coordinates by setting `has_coords = TRUE`, and specify limit equal
to 1000 to retrieve a maximum of 1000 occurrences.

```
library('spocc')
df <- occ(query = "Bradypus variegatus", from = "gbif",
          date = c("2000-01-01", "2019-12-31"),
          gbifopts = list(country = "CR"),
          has_coords = TRUE, limit = 1000)
d <- occ2df(df)
```

Then, we transform the point data to a `sf` object with `st_as_sf()`, assign
the coordinate reference system given by the EPSG code 4326 to represent
longitude and latitude coordinates. Figure 6.7 shows the retrieved locations of
sloths in Costa Rica.

```
library(sf)
d <- st_as_sf(d, coords = c("longitude", "latitude"))
st_crs(d) <- 4326
mapview(d)
```

FIGURE 6.7: Locations of sloths in Costa Rica obtained with **spocc**.

6.8   Population, health, and other spatial data

Other packages that can be used to obtain spatial data for the world or specific countries include the following: The **wopr** package (Leasure et al., 2023) provides access to the WorldPop Open Population Repository[12] and provides estimates of population sizes for specific geographic areas. These data are collected by the WorldPop Hub (https://hub.worldpop.org/), which provides open high-resolution geospatial data on population count and density, demographic and dynamics, with a focus on low- and middle-income countries.

The **rdhs** package (Watson and Eaton, 2022) gives the users the ability to access and make analysis on the Demographic and Health Survey (DHS)[13] data. The **malariaAtlas** package (Pfeffer et al., 2020) can be used to download, visualize and manipulate global malaria data hosted by the Malaria Atlas Project[14]. The **openair** package (Carslaw et al., 2023) allows us to obtain air quality data and other atmospheric composition data.

Many other spatial datasets are included in several packages mainly to demonstrate the packages' functionality. For example, **spatstat** (Baddeley et al., 2022) contains point pattern data that can be listed with `data(package="spatstat.data")`. The **spData** package (Bivand et al., 2022) also includes diverse spatial datasets that can be used for teaching spatial data analysis.

R also has packages that allow us to geocode place names or addresses. For example, the packages **ggmap** (Kahle et al., 2022) and **opencage** (Possenriede et al., 2021) can be used to convert names to geographic coordinates.

---

[12]https://wopr.worldpop.org/
[13]https://dhsprogram.com/
[14]https://malariaatlas.org/

Part II

Areal data

7

Spatial neighborhood matrices

Areal or lattice data arise when a study region is partitioned into a finite number of areas at which outcomes are aggregated. Examples of areal data are the number of individuals with a certain disease in municipalities of a country, the number of road accidents in provinces, or the average housing prices in districts of a city.

The concept of spatial neighborhood is useful for the exploration of areal data to assess spatial autocorrelation and find out whether close areas have similar or dissimilar values. Spatial neighbors can be defined in several ways depending on the variable of interest and the specific setting. The simplest neighborhood definition assumes that neighbors are areas that share a common border, perhaps a vertex. We can also expand the idea of neighborhood to include areas that are close, but not necessarily adjacent, by assuming neighbors are areas that are within some distance apart.

Given a spatial neighborhood definition, we can construct a spatial neighborhood matrix which will allow us to assess spatial autocorrelation. The elements of the spatial neighborhood matrix can be viewed as weights that spatially connect areas. In this matrix, entries corresponding to close areas will have more weight than entries corresponding to areas that are farther apart.

The **spdep** package (Bivand, 2022) contains a number of functions to deal with spatial dependence structures. Some of its functions can be used to construct spatial neighborhood matrices and perform spatial autocorrelation analyses. For example, the functions `poly2nb()` and `dnearneigh()` can be used to create neighbor lists based on contiguity and distance criteria, respectively. Spatial neighborhood matrices can be built from the neighbor lists using the `nb2listw()` function.

In this chapter, we demonstrate how to use the **spdep** package to construct several types of spatial neighborhood structures and matrices using the map of the 49 neighborhoods of Columbus, Ohio, USA. We read the map which is in the `columbus` shapefile of the **spData** package and assign it to a variable called `map` as follows.

```
library(spData)
library(sf)
```

```
library(spdep)
library(ggplot2)
map <- st_read(system.file("shapes/columbus.shp",
                 package = "spData"), quiet = TRUE)
```

7.1   Neighbors based on contiguity

Neighbors based on contiguity are constructed by assuming that neighbors of
a given area are other areas that share a common boundary. Figure 7.1 shows
two types of contiguity neighbors. Neighbors can be of type Queen if a single
shared boundary point meets the contiguity condition, or Rook if more than
one shared point is required to meet the contiguity condition.

Queen   **Rook**

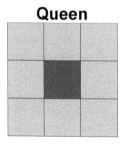   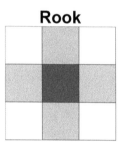

FIGURE 7.1: Neighbors based on contiguity. Area of interest is represented
in black and its neighbors in gray.

The function `poly2nb()` of the **spdep** package can be used to construct a
list of neighbors based on areas with contiguous boundaries, that is, areas
sharing one or more boundary point. `poly2nb()` accepts a list of polygons and
returns a list of class nb with the neighbors of each area. The default type
in `poly2nb()` is queen = TRUE so neighbors of a given area are other areas
sharing a common point or more than one point. Here, we use `poly2nb()` to
calculate the neighbors of each of the regions of Columbus based on Queen
contiguity.

```
library(spdep)
nb <- spdep::poly2nb(map, queen = TRUE)
head(nb)
```

```
[[1]]
[1] 2 3

[[2]]
[1] 1 3 4

[[3]]
[1] 1 2 4 5

[[4]]
[1] 2 3 5 8

[[5]]
[1]  3  4  6  8  9 11 15 16

[[6]]
[1] 5 9
```

Figure 7.2 shows a map with the neighbors obtained. This plot is obtained by first plotting the map, and then overlapping the neighborhood structure with the `plot.nb()` function passing the neighbor list and the coordinates of the map.

```
plot(st_geometry(map), border = "lightgray")
plot.nb(nb, st_geometry(map), add = TRUE)
```

We can plot the neighbors of a given area by adding a new column in `map` representing the neighbors of the area. For example, Figure 7.2 shows the neighbors of area 20.

```
id <- 20 # area id
map$neighbors <- "other"
map$neighbors[id] <- "area"
map$neighbors[nb[[id]]] <- "neighbors"
ggplot(map) + geom_sf(aes(fill = neighbors)) + theme_bw() +
  scale_fill_manual(values = c("gray30", "gray", "white"))
```

Given a neighbor list, the cardinality function `spdep::card()` counts the number of neighbors of each area. We can also obtain the number of neighbors of each area with `lengths(nb)`. Then, we can use `table(1:nrow(map), card(nb))` to build a table with the areas' ids in rows, and the number of neighbors in columns.

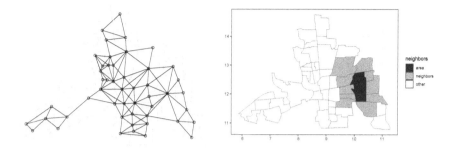

FIGURE 7.2: Left: Map of neighbors based on contiguity. Right: Map of neighbors of area 20 based on contiguity.

7.2   Neighbors based on k nearest neighbors

We can also consider as neighbors of an area its k nearest neighbors based on the distance separating them. For example, Figure 7.3 represents the 3 nearest neighbors of an area. The function `knearneigh()` of **spdep** allows us to obtain a matrix with the indices of points belonging to the set of the k nearest neighbors of each area. The arguments of this function include a matrix of point coordinates and the number `k` of nearest neighbors to be returned. Then, we can use `knn2nb()` to convert this list into a neighbor list of class `nb` with the integer vectors containing the ids of the neighbors. Figure 7.3 shows a map of the nearest neighbors of `map` with order $k = 3$.

```
# Neighbors based on 3 nearest neighbors
coo <- st_centroid(map)
nb <- knn2nb(knearneigh(coo, k = 3)) # k number nearest neighbors
plot(st_geometry(map), border = "lightgray")
plot.nb(nb, st_geometry(map), add = TRUE)
```

7.3   Neighbors based on distance

Neigborhood structures can also be defined by considering neighbors areas that are within a given distance (Figure 7.4). The `dnearneigh()` function of **spdep** builds a list of neighbors based on a distance between specific lower and upper bounds. The arguments of `dnearneigh()` include the object with the point coordinates, and the lower and upper distance bounds (`d1` and `d2`).

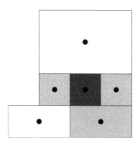 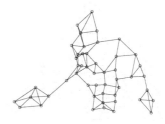

FIGURE 7.3: Left: Neighbors based on 3 nearest neighbors. Area of interest is represented in black and its neighbors in gray. Right: Map of neighbors based on 3 nearest neighbors.

For example, Figure 7.4 shows the map of neighbors obtained by considering neighbors areas separated by a distance less than 0.4.

```
# Neighbors based on distance
nb <- dnearneigh(x = st_centroid(map), d1 = 0, d2 = 0.4)
plot(st_geometry(map), border = "lightgray")
plot.nb(nb, st_geometry(map), add = TRUE)
```

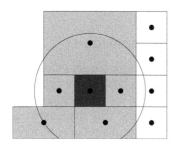 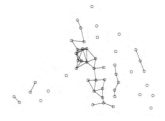

FIGURE 7.4: Left: Neighbors based on distance. Area of interest is represented in black and its neighbors in gray. Circle's center is the centroid of the area of interest, and circle's radius is the distance. Right: Map of neighbors separated by a distance less than 0.4.

Note that we can also determine an appropriate upper distance to ensure that each area has at least k neighbors. To determine an appropriate upper distance bound for a given number of neighbors k preferred, we can proceed as follows. First, we use the **spdep::knearneigh()** function to obtain the k nearest neighbors of each of the areas. We use this function passing a matrix of point coordinates and the number k of chosen nearest neighbors to obtain a matrix with the **k** nearest neighbors of each area. Then, we use `spdep::knn2nb()` to convert this list into a neighbor list of class **nb** with the ids of the neighbors. For example, here we determine the upper distance bound to ensure all areas have at least one neighbor.

```
coo <- st_centroid(map)
# k is the number nearest neighbors
nb1 <- knn2nb(knearneigh(coo, k = 1))
```

Then, we use `spdep::nbdists()` passing the **nb** list with the neighbors and the matrix of point coordinates to obtain the distances along the links. Finally, we can compute summaries to determine the upper distance bound.

```
dist1 <- nbdists(nb1, coo)
summary(unlist(dist1))
```

Min.	1st Qu.	Median	Mean	3rd Qu.	Max.
0.128	0.254	0.316	0.329	0.404	0.619

In this example, the maximum distance is 0.62 and we can take this value as an upper bound of the distance to ensure each area has at least one neighbor.

7.4   Neighbors of order k based on contiguity

Figure 7.5 shows first and second order neighbors based on contiguity. The `nblab()` function of the **spdep** package creates higher order neighbor lists, where higher order neighbors are lags links from each other on the graph described by the input neighbors list of class **nb**. The arguments of this function are a neighbors list of class **nb**, and the maximum lag considered. The returned object is a list of lagged neighbors lists each with class **nb**.

Here, we use `nblag()` with the maximum lag equal to `maxlag = 2` to create a list containing a list of neighbors of order 1, and a list of neighbors of order 2 (Figure 7.6).

```
library(spdep)
nb <- poly2nb(map, queen = TRUE)
nblags <- spdep::nblag(neighbours = nb, maxlag = 2)
```

Rook. First order

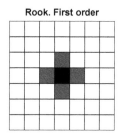

Queen. First order

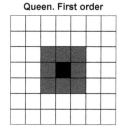

Rook. Second order

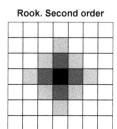

Queen. Second order

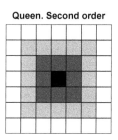

FIGURE 7.5: Rook and Queen neighbors of first (dark gray) and second (light gray) order.

```
# Neighbors of first order
plot(st_geometry(map), border = "lightgray")
plot.nb(nblags[[1]], st_geometry(map), add = TRUE)
```

```
# Neighbors of second order
plot(st_geometry(map), border = "lightgray")
plot.nb(nblags[[2]], st_geometry(map), add = TRUE)
```

The function `nblag_cumul()` of **spdep** cumulates a list of lagged neighbors as the output of `nblag()` and returns a single neighbor list of class **nb** containing the neighbors of order 1 until the maximum lag considered. Figure 7.6 shows the map of neighbors of order 1 until order 2.

```
# Neighbors of order 1 until order 2
nb <- spdep::poly2nb(map, queen = TRUE)
nblagsc <- spdep::nblag_cumul(nblags)
plot(st_geometry(map), border = "lightgray")
plot.nb(nblagsc, st_geometry(map), add = TRUE)
```

FIGURE 7.6: Map of neighbors based on contiguity. Neighbors of first order (left), second order (middle), and first order until second order (right).

7.5 Neighborhood matrices

A spatial neighborhood matrix W defines a neighborhood structure over the entire study region, and its elements can be viewed as weights. The (i,j)th element of W, denoted by w_{ij}, spatially connects areas i and j in some fashion. More weight is associated with areas closer to i than those farther away from i.

If neighbors are based on contiguity, we can construct a binary spatial matrix with $w_{ij} = 1$ if regions i and j share a common boundary, and $w_{ij} = 0$ otherwise. Customarily, w_{ii} is set to 0 for $i = 1, \ldots, n$. An example of binary spatial matrix is shown in Figure 7.7. Note that this choice of proximity measure results in a symmetric spatial matrix.

	A	B	C	D	E	Sum
A	0	1	1	1	0	3
B	1	0	1	1	1	4
C	1	1	0	1	0	3
D	1	1	1	0	1	4
E	0	1	0	1	0	2

Areas of the study region: A B / C D E

FIGURE 7.7: Left: Areas of the study region. Right: Spatial weight matrix calculated by assuming neighboring areas share a common boundary, and sum of weights for each area.

Other spatial weight definitions could be to use $w_{ij} = 1$ for all i and j within a specified distance, or to use $w_{ij} = 1$ if j is one of the k nearest neighbors of i. Weights w_{ij} can also be defined as the inverse distance between areas.

In addition, we may want to adjust for the total number of neighbors in each area and use a standardized matrix with entries $w_{std,i,j} = w_{ij}/\sum_{j=1}^{n} w_{ij}$. Note that in most situations, this matrix is not symmetric when the areas are irregularly shaped.

Spatial weights matrix based on a binary neighbor list

The function `nb2listw()` of the **spdep** package can be used to construct a spatial neighborhood matrix containing the spatial weights corresponding to a neighbors list. The neighbors can be binary or based on inverse distance values. To compute a spatial weights matrix based on a binary neighbor list, we use the `nb2listw()` function with the following arguments:

- `nb` list with neighbors,
- `style` indicates the coding scheme chosen. For example, `style = B` is the basic binary coding, and `W` is row standardized (1/number of neighbors),
- `zero.policy` is used to take into account regions with 0 neighbors. Specifically, `zero.policy = TRUE` permits the weight list to contain zero-length weights vectors, and `zero.policy = FALSE` stops the function with an error if there are empty neighbor sets.

```
nb <- poly2nb(map, queen = TRUE)
nbw <- spdep::nb2listw(nb, style = "W")
nbw$weights[1:3]
```

```
[[1]]
[1] 0.5 0.5

[[2]]
[1] 0.3333 0.3333 0.3333

[[3]]
[1] 0.25 0.25 0.25 0.25
```

We can visualize the spatial weight matrix by creating a matrix with the weights with `listw2mat()`, and using `lattice::levelplot()` to create the plot (Figure 7.8).

```
m1 <- listw2mat(nbw)
lattice::levelplot(t(m1),
scales = list(y = list(at = c(10, 20, 30, 40),
                       labels = c(10 20, 30, 40))))
```

Spatial weights matrix based on inverse distance values

Given a list of neighbors, we can use `nbdists()` to compute the distances
along the links. Then, we can construct the list with spatial weights based on
inverse distance values using `nb2listw()` where the argument `glist` is equal
to a list of general weights corresponding to neighbors. Figure 7.8 shows the
spatial weight matrix obtained.

```
coo <- st_centroid(map)
nb <- poly2nb(map, queen = TRUE)
dists <- nbdists(nb, coo)
ids <- lapply(dists, function(x){1/x})


nbw <- nb2listw(nb, glist = ids, style = "B")
nbw$weights[1:3]

[[1]]
[1] 1.670 1.725

[[2]]
[1] 1.670 1.405 2.943

[[3]]
[1] 1.725 1.405 1.783 1.911


m2 <- listw2mat(nbw)
lattice::levelplot(t(m2),
scales = list(y = list(at = c(10, 20, 30, 40),
                       labels = c(10, 20, 30, 40)))))
```

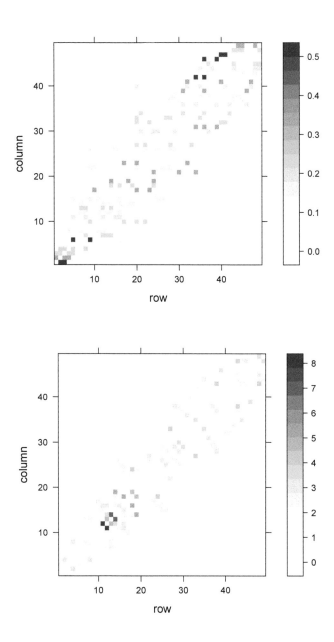

FIGURE 7.8: Spatial weights matrix based on a binary neighbor list (top), and inverse distance values (bottom).

8

Spatial autocorrelation

Spatial autocorrelation is used to describe the extent to which a variable is correlated with itself through space. This concept is closely related to Tobler's First Law of Geography, which states that "everything is related to everything else, but near things are more related than distant things" (Tobler, 1970). Positive spatial autocorrelation occurs when observations with similar values are closer together (i.e., clustered). Negative spatial autocorrelation occurs when observations with dissimilar values are closer together (i.e., dispersed). Figure 8.1 shows three configurations of areas showing different types of spatial autocorrelation.

Negative spatial autocorrelation	No spatial autocorrelation	Positive spatial autocorrelation
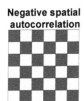	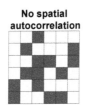	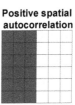

FIGURE 8.1: Examples of configurations of areas showing different types of spatial autocorrelation.

Spatial autocorrelation can be assessed using indices that summarize the degree to which similar observations tend to occur near each other over the study area. Two common indices that are used to assess spatial autocorrelation in areal data are Moran's I (Moran, 1950) and Geary's C (Geary, 1954).

In this chapter, we use the Moran's I to test the spatial autocorrelation of housing prices in 506 census tracts in Boston, USA, in 1978 obtained from the **spData** package (Bivand et al., 2022). The data has a variable called MEDV with the median prices of owner-occupied housing in $1000 USD. We create the variable vble with the values of MEDV that will be used in the analysis. Figure 8.2 shows the map created with the housing prices using **mapview** (Appelhans et al., 2022).

```
library(spData)
library(sf)
```

```
library(mapview)
map <- st_read(system.file("shapes/boston_tracts.shp",
                           package = "spData"), quiet = TRUE)
map$vble <- map$MEDV
mapview(map, zcol = "vble")
```

FIGURE 8.2: Median prices of owner-occupied housing in $1000 USD in census tracts of Boston in 1978.

8.1   Global Moran's I

The Global Moran's I (Moran, 1950) takes the form

$$I = \frac{n \sum_i \sum_j w_{ij}(Y_i - \bar{Y})(Y_j - \bar{Y})}{(\sum_{i \neq j} w_{ij}) \sum_i (Y_i - \bar{Y})^2},$$

where n is the number of regions, Y_i is the observed value of the variable of interest in region i, and $\bar{Y}$ is the mean of all values. w_{ij} are spatial weights that denote the spatial proximity between regions i and j, with $w_{ii} = 0$ and $i, j = 1, \ldots, n$. The definition of the spatial weights depends on the variable of study and the specific setting.

We can test the presence of spatial autocorrelation using the Moran's I, which quantifies how similar each region is with its neighbors and averages all these assessments. Under the null hypothesis of no spatial autocorrelation, observations Y_i are independent identically distributed, and I is asymptotically normally distributed with mean and variance equal to

$$E[I] = \frac{-1}{n-1}$$

and

$$Var[I] = \frac{n^2(n-1)S_1 - n(n-1)S_2 - 2S_0^2}{(n+1)(n-1)^2 S_0^2},$$

where

$$S_0 = \sum_{i \neq j} w_{ij}, \quad S_1 = \frac{1}{2} \sum_{i \neq j} (w_{ij} + w_{ji})^2 \text{ and } S_2 = \sum_k \left(\sum_j w_{kj} + \sum_i w_{ik} \right)^2.$$

Moran's I values usually range from –1 to 1. Moran's I values significantly above $E[I] = -1/(n-1)$ indicate positive spatial autocorrelation or clustering. This occurs when neighboring regions tend to have similar values. Moran's I values significantly below $E[I]$ indicate negative spatial autocorrelation or dispersion. This happens when regions that are close to one another tend to have different values. Finally, Moran's I values around $E[I]$ indicate randomness, that is, absence of spatial pattern.

When the number of regions is sufficiently large, I has a normal distribution and we can assess whether any given pattern deviates significantly from a random pattern by comparing the z-score

$$z = \frac{I - E(I)}{Var(I)^{1/2}}$$

to the standard normal distribution. An alternative approach to judge significance is Monte Carlo randomization. This method creates random patterns by reassigning the observed values among the areas and calculates the Moran's I for each of the patterns, providing a randomization distribution for the Moran's I. If the observed value of Moran's I lies in the tails of this distribution, the assumption of independence among observations is rejected. Thus, we can test spatial autocorrelation by following these steps:

1. State the null and alternative hypotheses:
    $H_0 : I = E[I]$ (no spatial autocorrelation),
    $H_1 : I \neq E[I]$ (spatial autocorrelation).

2. Choose the significance level α we are willing to tolerate, which represents the maximum value for the probability of incorrectly rejecting the null hypothesis when it is true (usually $\alpha = 0.05$).

3. Calculate the test statistic:

    $$z = \frac{I - E(I)}{Var(I)^{1/2}}.$$

4. Find the p-value for the observed data by comparing the z-score to the standard normal distribution or via Monte Carlo randomization. The p-value is the probability of obtaining a test statistic as extreme as or more extreme than the one observed test statistic in the direction of the alternative hypothesis, assuming the null hypothesis is true.

5. Make one of these two decisions and state a conclusion:
    If p-value $< \alpha$, we reject the null hypothesis. We conclude data provide evidence for the alternative hypothesis.
    If p-value $\geq \alpha$, we fail to reject the null hypothesis. The data do not provide evidence for the alternative hypothesis.

8.2   The `moran.test()` function

The function `moran.test()` of the **spdep** package can be used to test spatial autocorrelation using Moran's I. The arguments of `moran.test()` are a numeric vector with the data, a list with the spatial weights, and the type of hypothesis. The argument that denotes the hypothesis is called `alternative` and can be set equal to `greater` (default), `less` or `two.sided` to represent a different alternative hypothesis. In this example, we specify the null and alternative hypothesis as follows:

$H_0 : I \leq E[I]$ (negative spatial autocorrelation or no spatial autocorrelation),
$H_1 : I > E[I]$ (positive spatial autocorrelation).

We use `moran.test()` to test this hypothesis by setting `alternative` = `"greater"`. The list with the spatial weights is calculated by first obtaining the neighbors of each area with the `poly2nb()` function, and then creating a list containing the neighbors with the `nb2listw()` function of **spdep**.

```
# Neighbors
library(spdep)
nb <- poly2nb(map, queen = TRUE) # queen shares point or border
nbw <- nb2listw(nb, style = "W")

# Global Moran's I
gmoran <- moran.test(map$vble, nbw,
                     alternative = "greater")
gmoran
```

```
    Moran I test under randomisation

data:  map$vble
weights: nbw

Moran I statistic standard deviate = 23, p-value
<2e-16
alternative hypothesis: greater
sample estimates:
Moran I statistic      Expectation          Variance
      0.6266754         -0.0019802         0.0007249
```

```
gmoran[["estimate"]][["Moran I statistic"]] # Moran's I
```

```
[1] 0.6267
```

```
gmoran[["statistic"]] # z-score
```

```
Moran I statistic standard deviate
                             23.35
```

```
gmoran[["p.value"]] # p-value
```

```
[1] 6.923e-121
```

The object returned by moran.test() provides the Moran's I statistic, the z-score and the p-value. We observe the p-value obtained is lower than the significance level 0.05. Then, we reject the null hypothesis and conclude there is evidence for positive spatial autocorrelation.

The same conclusion is obtained if we use a Monte Carlo approach to assess significance. This approach creates random patterns by reassigning the values among the fixed areas and calculates the Moran's *I* for each of these patterns. Then, it calculates the p-value as the proportion of values as extreme or more extreme than the statistic observed in the direction of the alternative hypothesis. Here, we conduct a Monte Carlo randomization approach using the `moran.mc()` function setting the number of simulations to `nsim = 999`. Figure 8.3 shows the histogram of the Moran's *I* values for each of the simulated patterns, as well as the Moran's *I* obtained with the real data. We observe a p-value lower than 0.05 indicating that the data presents positive spatial autocorrelation.

```
gmoranMC <- moran.mc(map$vble, nbw, nsim = 999)
gmoranMC
```

```
    Monte-Carlo simulation of Moran I

data:  map$vble
weights: nbw
number of simulations + 1: 1000

statistic = 0.63, observed rank = 1000, p-value
= 0.001
alternative hypothesis: greater
```

```
hist(gmoranMC$res)
abline(v = gmoranMC$statistic, col = "red")
```

8.3   Moran's *I* scatterplot

The `moran.plot()` function can be used to construct a Moran's *I* scatterplot to visualize the spatial autocorrelation in the data. This plot displays the observations of each area against its spatially lagged values. The spatially lagged value for a given area is calculated as a weighted average of the neighboring values for that area. This value can be computed, for example, using the `lag.listw()` function of **spdep** passing the values and the spatial weights corresponding to the areas. Figure 8.4 shows the Moran's *I* scatterplot for the housing prices data obtained with the `moran.plot()` function passing the housing prices and the spatial weights corresponding to the Boston tracts. We observe a positive linear relationship between the observations and their spatially lagged values. Using this plot, we can also identify data points that

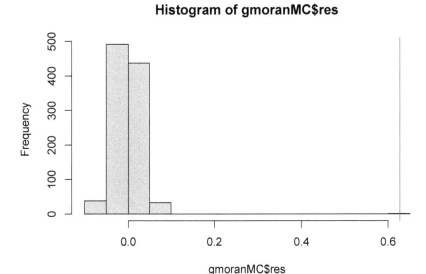

FIGURE 8.3: Histogram of the Moran's I values for each of the simulated patterns in the Monte Carlo randomization approach. The red line represents the Moran's I obtained for the real data.

have a high influence on the linear relationship between the data and the lagged values.

```
moran.plot(map$vble, nbw)
```

8.4 Local Moran's I

We have seen that the Global Moran's I provides an index to assess the spatial autocorrelation for the whole study region. There is often interest in providing a local measure of similarity between each area's value and those of nearby areas. Local Indicators of Spatial Association (LISA) (Anselin, 1995) are designed to provide an indication of the extent of significant spatial clustering of similar values around each observation. A desirable property is that the sum of the LISA's values across all regions is equal to a multiple of the global indicator of spatial association. As a result, global statistics may be decomposed into a set

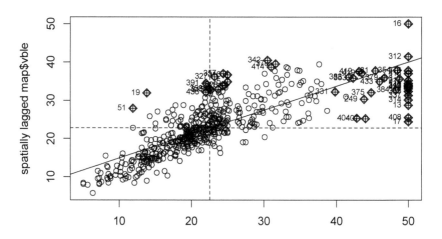

FIGURE 8.4: Moran's I scatterplot showing the observations against its spatially lagged values.

of local statistics and most LISAs are defined as local versions of well-known global indexes.

One of the most popular LISAs is the local version of Moran's I. For the ith region, the local Moran's I defined as

$$I_i = \frac{n(Y_i - \bar{Y})}{\sum_j (Y_j - \bar{Y})^2} \sum_j w_{ij}(Y_j - \bar{Y}).$$

Note that the global Moran's I is proportional to the sum of the local Moran's I obtained for all regions:

$$I = \frac{1}{\sum_{i \neq j} w_{ij}} \sum_i I_i.$$

Typically, the values of the LISAs are mapped to indicate the location of areas with comparatively high or low local association with neighboring areas. A high value for I_i suggests that the area is surrounded by areas with similar values. Such an area is part of a cluster of high observations, low observations, or moderate observations. A low value for I_i indicates that the area is surrounded by areas with dissimilar values. Such an area is an outlier indicating that the observation of area i is different from most or all of the observations of its neighbors.

To interpret the local Moran's I for each of the areas, it is necessary to compute a map of p-values representing the probability of exceeding the observed values assuming the null hypothesis is true. These p-values, regardless of the presence or absence of global spatial association, may be obtained by a simulation process with a conditional randomization approach. In this approach, the observed value Y_i at region i is fixed, and the remaining values are randomly reassigned over the other regions.

8.5 The localmoran() function

The localmoran() function of the **spdep** package can be used to compute the Local Moran's I for a given dataset. The arguments of localmoran() include a numeric vector with the values of the variable, a list with the neighbor weights, and the name of an alternative hypothesis that can be set equal to greater (default), less or two.sided. The returned object of the localmoran() function contains the following information:

- Ii: Local Moran's I statistic for each area,
- E.Ii: Expectation Local Moran's I statistic,
- Var.Ii: Variance Local Moran's I statistic,
- Z.Ii: z-score,
- Pr(z > E(Ii)), Pr(z < E(Ii)) or Pr(z != E(Ii)): p-value for an alternative hypothesis greater, less or two.sided, respectively.

Here, we use the localmoran() function to compute the Local Moran's I for the housing prices data. We set alternative = "greater" which corresponds to testing H_0: no or negative spatial autocorrelation vs. H_1: positive spatial autocorrelation.

```
lmoran <- localmoran(map$vble, nbw, alternative = "greater")
head(lmoran)
```

```
          Ii        E.Ii      Var.Ii    Z.Ii
1 -0.3457508 -5.254e-04 3.275e-02 -1.9075
2  0.0175875 -1.627e-05 2.046e-03  0.3892
3  0.0123380 -6.557e-07 4.090e-05  1.9294
4 -0.0001654 -1.059e-07 1.332e-05 -0.0453
5  0.3591629 -1.428e-04 7.899e-03  4.0428
6  0.0545611 -1.626e-04 1.357e-02  0.4697
  Pr(z > E(Ii))
```

```
1       9.718e-01
2       3.486e-01
3       2.684e-02
4       5.181e-01
5       2.641e-05
6       3.193e-01
```

Figure 8.5 depicts maps created with **tmap** (Tennekes, 2022) showing the
housing prices, the local Moran's I, z-scores, and p-values. Areas with p-value
less than the significance level 0.05 (or with z-scores higher than `qnorm(0.95)`
= 1.65) correspond to areas for which we would reject the null hypothesis and
conclude they present positive spatial autocorrelation.

```
library(tmap)
tmap_mode("plot")


map$lmI <- lmoran[, "Ii"] # local Moran's I
map$lmZ <- lmoran[, "Z.Ii"] # z-scores
# p-values corresponding to alternative greater
map$lmp <- lmoran[, "Pr(z > E(Ii))"]

p1 <- tm_shape(map) +
  tm_polygons(col = "vble", title = "vble", style = "quantile") +
  tm_layout(legend.outside = TRUE)

p2 <- tm_shape(map) +
  tm_polygons(col = "lmI", title = "Local Moran's I",
              style = "quantile") +
  tm_layout(legend.outside = TRUE)

p3 <- tm_shape(map) +
  tm_polygons(col = "lmZ", title = "Z-score",
              breaks = c(-Inf, 1.65, Inf)) +
  tm_layout(legend.outside = TRUE)

p4 <- tm_shape(map) +
  tm_polygons(col = "lmp", title = "p-value",
              breaks = c(-Inf, 0.05, Inf)) +
  tm_layout(legend.outside = TRUE)

tmap_arrange(p1, p2, p3, p4)
```

If we used `alternative = "two.sided"` instead of `alternative =`
`"greater"`, we would be testing H_0: no spatial autocorrelation vs. H_1: positive

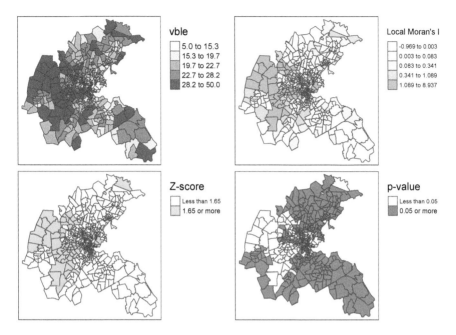

FIGURE 8.5: Boston housing prices, local Moran's I, z-scores, and p-values.

or negative spatial autocorrelation. In this two-sided test, z-score values lower than −1.96 indicate negative spatial autocorrelation, and z-score values greater than 1.96 indicate positive spatial autocorrelation. Figure 8.6 features a map showing the areas with negative, no, and positive spatial autocorrelation, obtained by breaking the legend according to these z-score values.

```
tm_shape(map) + tm_polygons(col = "lmZ",
title = "Local Moran's I", style = "fixed",
breaks = c(-Inf, -1.96, 1.96, Inf),
labels = c("Negative SAC", "No SAC", "Positive SAC"),
palette =  c("blue", "white", "red")) +
tm_layout(legend.outside = TRUE)
```

8.6   Clusters

Local Moran's I allows us to identify clusters of the following types:

- High-High: areas of high values with neighbors of high values,
- High-Low: areas of high values with neighbors of low values,

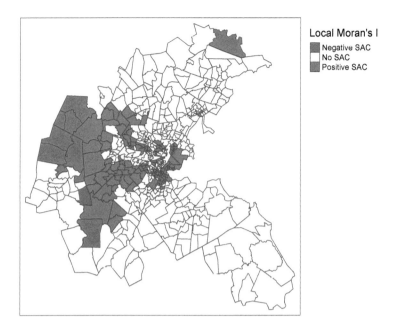

FIGURE 8.6: Boston areas showing negative, no, and positive spatial auto-correlation according to the local Moran's I.

- Low-High: areas of low values with neighbors of high values,
- Low-Low: areas of low values with neighbors of low values.

To detect clusters, we first use the `localmoran()` function to calculate the local Moran's I. The p-values for the alternative hypothesis `"two.sided"` are in column 5 of the returned object.

```
lmoran <- localmoran(map$vble, nbw, alternative = "two.sided")
head(lmoran)
```

```
           Ii        E.Ii     Var.Ii     Z.Ii
1 -0.3457508 -5.254e-04  3.275e-02  -1.9075
2  0.0175875 -1.627e-05  2.046e-03   0.3892
3  0.0123380 -6.557e-07  4.090e-05   1.9294
4 -0.0001654 -1.059e-07  1.332e-05  -0.0453
5  0.3591629 -1.428e-04  7.899e-03   4.0428
6  0.0545611 -1.626e-04  1.357e-02   0.4697
  Pr(z != E(Ii))
1       5.645e-02
2       6.971e-01
3       5.368e-02
```

```
4       9.639e-01
5       5.282e-05
6       6.386e-01
```

```
map$lmp <- lmoran[, 5] # p-values are in column 5
```

Then, we identify the clusters of each type by using the information provided by the Moran's *I* scatterplot obtained with the `moran.plot()` function (Figure 8.7).

```
mp <- moran.plot(as.vector(scale(map$vble)), nbw)
```

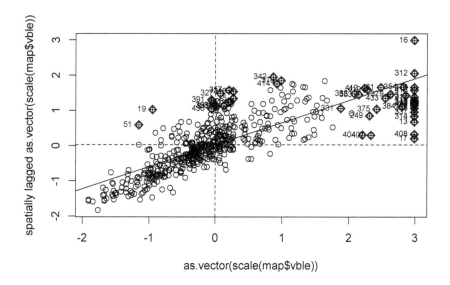

FIGURE 8.7: Moran's *I* scatterplot showing the scaled values against its spatially lagged values.

```
head(mp)
```

```
           x       wx is_inf labels      dfb.1_
1 -0.514597  0.67056  FALSE      1    0.090885
2 -0.090551 -0.19384  FALSE      2   -0.015162
3  0.018179  0.67735  FALSE      3    0.060002
4  0.007306 -0.02259  FALSE      4   -0.004886
```

```
5   0.268258  1.33623    TRUE      5   0.107458
6  -0.286265 -0.19022   FALSE      6  -0.003359
        dfb.x      dffit   cov.r     cook.d       hat
1 -4.682e-02  0.102234 0.9900 5.193e-03 0.002501
2  1.374e-03 -0.015224 1.0055 1.161e-04 0.001993
3  1.092e-03  0.060012 0.9987 1.798e-03 0.001977
4 -3.573e-05 -0.004886 1.0059 1.196e-05 0.001976
5  2.885e-02  0.111265 0.9832 6.131e-03 0.002119
6  9.624e-04 -0.003494 1.0061 6.115e-06 0.002139
```

Specifically, we identify the cluster types by using the quadrants of the scaled
values (mp$x) and their spatially lagged values (mp$wx), and the p-values
obtained with the local Moran's I for each of the areas (map$lmp). The
classification of the clusters is as follows. Areas with significant local Moran's
I are classified as high-high if both the value and its corresponding spatially
lagged value are positive, low-low if both the value and its spatially lagged
value are negative, high-low if the the value is positive and the spatially lagged
value negative, and low-high is the value is negative and the spatially lagged
value positive.

We create the variable quadrant denoting the type of cluster for each of the
areas using the quadrant corresponding to its value and its spatially lagged
value, and the p-value. Specifically, areas with quadrant equal to 1, 2, 3, and
4 correspond to clusters of type high-high, low-low, high-low, and low-high,
respectively. Areas with quadrant equal to 5 are non-significant.

```
map$quadrant <- NA
# high-high
map[(mp$x >= 0 & mp$wx >= 0) & (map$lmp <= 0.05), "quadrant"]<- 1
# low-low
map[(mp$x <= 0 & mp$wx <= 0) & (map$lmp <= 0.05), "quadrant"]<- 2
# high-low
map[(mp$x >= 0 & mp$wx <= 0) & (map$lmp <= 0.05), "quadrant"]<- 3
# low-high
map[(mp$x <= 0 & mp$wx >= 0) & (map$lmp <= 0.05), "quadrant"]<- 4
# non-significant
map[(map$lmp > 0.05), "quadrant"] <- 5
```

Figure 8.8 shows the map of the clusters obtained with the Boston housing
prices data.

```
tm_shape(map) + tm_fill(col = "quadrant", title = "",
breaks = c(1, 2, 3, 4, 5, 6),
palette =  c("red", "blue", "lightpink", "skyblue2", "white"),
labels = c("High-High", "Low-Low", "High-Low",
```

```
            "Low-High", "Non-significant")) +
tm_legend(text.size = 1)  + tm_borders(alpha = 0.5) +
tm_layout(frame = FALSE,  title = "Clusters")  +
tm_layout(legend.outside = TRUE)
```

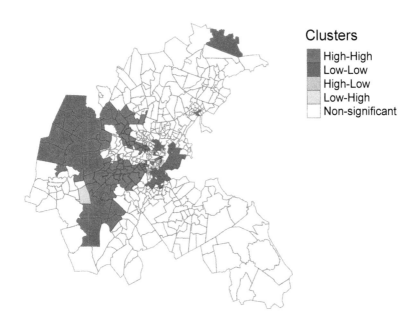

FIGURE 8.8: High-high, low-low, high-low, and low-high clusters detected in the Boston housing prices data.

9

Bayesian spatial models

Bayesian hierarchical models (Banerjee et al., 2004) can be used to analyze areal data that arise when an outcome variable is aggregated into areas that form a partition of the study region. Models can be specified to describe the variability in the response variable as a function of a number of covariates known to affect the outcome, as well as random effects to model residual variation not explained by the covariates. This provides a flexible and robust approach that allows us to assess the effects of explanatory variables, accommodate spatial autocorrelation, and quantify the uncertainty in the estimates obtained.

A commonly used spatial model is the Besag-York-Mollié (BYM) model (Besag et al., 1991). Let us assume that Y_i are observed outcomes at regions $i = 1, \ldots, n$ that can be modeled using a Normal distribution. The BYM model is specified as

$$Y_i \sim Normal(\mu_i, \sigma^2), i = 1, \ldots, n,$$

$$\mu_i = z_i \beta + u_i + v_i.$$

Here, the fixed effects $z_i \beta$ are expressed using a vector of intercept and p covariates corresponding to area i, $z_i = (1, z_{i1}, \ldots, z_{ip})$, and a coefficient vector $\beta = (\beta_0, \ldots, \beta_p)'$.

The model includes a spatial random effect u_i that accounts for the spatial dependence between outcomes indicating that areas that are close to each other may have similar values, and an unstructured exchangeable component v_i to model uncorrelated noise. The spatial random effect u_i can be modeled with an intrinsic conditional autoregressive model (CAR) that smooths the data according to a certain neighborhood structure. Specifically,

$$u_i | u_{-i} \sim N\left(\bar{u}_{\delta_i}, \frac{\sigma_u^2}{n_{\delta_i}}\right),$$

where $\bar{u}_{\delta_i} = n_{\delta_i}^{-1} \sum_{j \in \delta_i} u_j$, with δ_i and n_{δ_i} representing, respectively, the set of neighbors and the number of neighbors of area i. The unstructured component v_i is modeled as independent and identically distributed normal variables with zero mean and variance σ_v^2, $v_i \sim N(0, \sigma_v^2)$.

9.1   Bayesian inference with INLA

Bayesian hierarchical models can be fitted using a number of approaches such as integrated nested Laplace approximation (INLA) (Rue et al., 2009) and Markov chain Monte Carlo (MCMC) methods (Gelman et al., 2013). INLA is a computational approach to perform approximate Bayesian inference in latent Gaussian models. This includes a wide range of models such as generalized linear mixed models and spatial and spatio-temporal models. INLA uses a combination of analytical approximations and numerical integration to obtain approximated posterior distributions of the parameters and is very fast compared to MCMC methods.

The **R-INLA** package (Rue et al., 2022) can be used to fit models using INLA. The INLA website (http://www.r-inla.org) includes documentation, examples, and other resources about INLA and the **R-INLA** package, including books that provide an introduction to Bayesian data analysis using INLA as well as practical examples in a variety of settings (Wang et al., 2018; Krainski et al., 2019; Moraga, 2019; Gómez-Rubio, 2020).

To install **R-INLA**, we use the `install.packages()` function specifying the **R-INLA** repository since the package is not on CRAN.

```
install.packages("INLA",
repos = "https://inla.r-inla-download.org/R/stable", dep = TRUE)
library(INLA)
```

Then, to fit a model, we write the linear predictor as a formula object in R, and call the `inla()` function passing the formula, the family distribution, the data, and other options. The object returned by `inla()` contains the fitted model. This object can be inspected, and the posterior distributions can be post-processed using a set of functions provided by **R-INLA**. **R-INLA** also provides functionality to specify priors, as well as to obtain a number of criteria that allow us to assess and compare Bayesian models such as the deviance information criterion (DIC) (Spiegelhalter et al., 2002), the Watanabe-Akaike information criterion (WAIC) (Watanabe, 2010), and the conditional predictive ordinate (CPO) (Held et al., 2010).

9.2 Spatial modeling of housing prices

Here, we provide an example on how to specify and fit a Bayesian hierarchical model to estimate housing prices in Boston, Massachusetts, USA, using the **R-INLA** package.

9.2.1 Housing prices in Boston, Massachusetts, USA

The Boston housing prices are in the **spData** package (Bivand et al., 2022), and can be obtained with the `st_read()` function of the **sf** package (Pebesma, 2022a) as follows.

```
library(sf)
library(spData)
map <- st_read(system.file("shapes/boston_tracts.shp",
               package = "spData"), quiet = TRUE)
```

This dataset contains housing data of 506 Boston census tracts including median prices of owner-occupied housing in $1000 USD (`MEDV`), per capita crime (`CRIM`), and average number of rooms per dwelling (`RM`). We create the variable called `vble` with the logarithm of the median prices, and map this variable using **mapview** (Figure 9.1). The map suggests that the housing prices are greater in the west, and prices are related to those in neighboring areas.

```
library(mapview)
map$vble <- log(map$MEDV)
mapview(map, zcol = "vble")
```

We will model the logarithm of the median prices using as covariates the per capita crime (`CRIM`) and the average number of rooms per dwelling (`RM`). Figure 9.2 shows the relationships between pairs of variables visualized using the `ggpairs()` function of the **GGally** package (Schloerke et al., 2021). We observe a negative relationship between the logarithm of housing price and crime, and a positive relationship between the logarithm of housing price and the average number of rooms.

```
library(GGally)
ggpairs(data = map, columns = c("vble", "CRIM", "RM"))
```

FIGURE 9.1: Logarithm of housing prices in Boston per census tract from the **spData** package.

9.2.2 Model

Let Y_i be the logarithm of housing price of area i, $i = 1, \ldots, n$. We fit a BYM model that considers Y_i as the response variable, and crime and number of rooms as covariates:

$$Y_i \sim N(\mu_i, \sigma^2), \; i = 1, \ldots, n,$$

$$\mu_i = \beta_0 + \beta_1 \times \text{crime}_i + \beta_2 \times \text{rooms}_i + u_i + v_i.$$

Here, β_0 is the intercept, and β_1 and β_2 represent, respectively, the coefficients of the covariates crime and number of rooms. u_i is a spatially structured effect modeled with a CAR structure, $u_i | \mathbf{u_{-i}} \sim N(\bar{u}_{\delta_i}, \frac{\sigma_u^2}{n_{\delta_i}})$. v_i is an unstructured effect modeled as $v_i \sim N(0, \sigma_v^2)$.

9.2.3 Neighborhood matrix

In the model, the spatial random effect u_i needs to be specified using a neighborhood structure. Here, we assume two areas are neighbors if they share a common boundary, and we create a neighborhood structure using functions

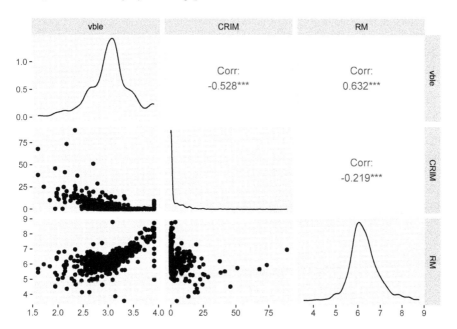

FIGURE 9.2: Relationship betwen the outcome variable logarithm of housing price (`vble`), and the covariates per capita crime (`CRIM`) and number of rooms (`RM`).

of the **spdep** package (Bivand, 2022). First, we use the `poly2nb()` function to create a neighbors list based on areas with contiguous boundaries. Each element of the list `nb` represents one area and contains the indices of its neighbors. For example, `nb[[1]]` contains the neighbors of area 1.

```
library(spdep)
library(INLA)
nb <- poly2nb(map)
head(nb)

[[1]]
[1]    2    3    6    8 311 313 314 369

[[2]]
[1] 1 3 4 6

[[3]]
[1]    1    2    4    5 369 371 375 376

[[4]]
```

```
[1] 2 3 5 6

[[5]]
[1]    3   4   6   7 375 376 411 413 418

[[6]]
[1] 1 2 4 5 7 8
```

Then, we use the `nb2INLA()` function to convert the `nb` list into a file called
`map.adj` with the representation of the neighborhood matrix as required by **R-INLA**. The `map.adj` file is saved in the working directory which can be obtained
with `getwd()`. Then, we read the `map.adj` file using the `inla.read.graph()`
function of **R-INLA**, and store it in the object `g` which we later use to specify
the spatial model using **R-INLA**.

```
nb2INLA("map.adj", nb)
g <- inla.read.graph(filename = "map.adj")
```

9.2.4   Model formula and `inla()` call

We specify the model formula by including the response variable, the ~ symbol,
and the fixed and random effects. By default, there is an intercept so we
do not need to include it in the formula. In the formula, random effects are
specified with the `f()` function. This function includes as first argument an
index vector that specifies the element of the random effect that applies to
each observation, and as second argument the model name. For the spatial
random effect u_i, we use `model = "besag"` with neighborhood matrix given by
`g`. The option `scale.model = TRUE` is used to make the precision parameter
of models with different CAR priors comparable (Freni-Sterrantino et al.,
2018). For the unstructured effect v_i, we choose `model = "iid"`. The index
vectors of the random effects are given by `re_u` and `re_v` created for u_i and
v_i, respectively. These vectors are equal to $1, \ldots, n$, where n is the number of
areas.

```
map$re_u <- 1:nrow(map)
map$re_v <- 1:nrow(map)


formula <- vble ~ CRIM + RM +
  f(re_u, model = "besag", graph = g, scale.model = TRUE) +
  f(re_v, model = "iid")
```

Note that in **R-INLA**, the BYM model can also be specified with `model
= "bym"` and this comprises both the spatial and unstructured components.

Alternatively, we can use the BYM2 model (Simpson et al., 2017) which is a new parametrization of the BYM model that uses a scaled spatial component $\boldsymbol{u}_*$ and an unstructured component $\boldsymbol{v}_*$:

$$\boldsymbol{b} = \frac{1}{\sqrt{\tau_b}}(\sqrt{1-\phi}\boldsymbol{v}_* + \sqrt{\phi}\boldsymbol{u}_*).$$

In this model, the precision parameter $\tau_b > 0$ controls the marginal variance contribution of the weighted sum of $\boldsymbol{u}_*$ and $\boldsymbol{v}_*$. The mixing parameter $0 \le \phi \le 1$ measures the proportion of the marginal variance explained by the spatial component $\boldsymbol{u}_*$. Thus, the BYM2 model is equal to an only spatial model when $\phi = 1$, and an only unstructured spatial noise when $\phi = 0$ (Riebler et al., 2016). The formula of the model using the BYM2 component is specified as follows.

```
formula <- vble ~ CRIM + RM + f(re_u, model = "bym2", graph = g)
```

Then, we fit the model by calling the `inla()` function specifying the formula, the family, the data, and using the default priors in **R-INLA**. We also set `control.predictor = list(compute = TRUE)` and `control.compute = list(return.marginals.predictor = TRUE)` to compute and return the posterior means of the predictors.

```
res <- inla(formula, family = "gaussian", data = map,
control.predictor = list(compute = TRUE),
control.compute = list(return.marginals.predictor = TRUE))
```

9.2.5 Results

The resulting object `res` contains the fit of the model. We can use `summary(res)` to obtain a summary of the fitted model. `res$summary.fixed` contains a summary of the fixed effects.

```
res$summary.fixed
```

	mean	sd	0.025quant	0.5quant
(Intercept)	1.426736	0.088581	1.25290	1.426759
CRIM	-0.007846	0.001293	-0.01038	-0.007846
RM	0.260338	0.014051	0.23278	0.260334
	0.975quant	mode	kld	
(Intercept)	1.60044	1.426759	2.148e-10	
CRIM	-0.00531	-0.007846	2.193e-10	
RM	0.28791	0.260334	2.156e-10	

We observe the intercept $\hat{\beta}_0 = 1.427$ with a 95% credible interval equal to (1.253, 1.6). We observe the intercept $\hat{\beta}_0 = -0.008$ with a 95% credible interval equal to (−0.01, −0.005). This indicates crime is significantly negatively related to housing price. The coefficient of rooms is $\hat{\beta}_2 = 0.26$ with a 95% credible interval equal to (0.233, 0.288) indicating number of rooms is significantly positively related to housing price. Thus, the results suggest both crime and number of rooms are important in explaining the spatial pattern of housing prices.

We can type res$summary.fitted.values to obtain a summary of the posterior distributions of the response μ_i for each of the areas. Column mean indicates the posterior mean, and columns 0.025quant and 0.975quant are, respectively, the lower and upper limits of 95% credible intervals representing the uncertainty of the estimates obtained.

```
summary(res$summary.fitted.values)
```

```
      mean                 sd                0.025quant
Min.    :1.61    Min.    :0.00992    Min.    :1.59
1st Qu.:2.83    1st Qu.:0.00995    1st Qu.:2.81
Median :3.05    Median :0.00996    Median :3.03
Mean    :3.04    Mean    :0.00997    Mean    :3.01
3rd Qu.:3.22    3rd Qu.:0.00997    3rd Qu.:3.20
Max.    :3.91    Max.    :0.01121    Max.    :3.89
   0.5quant              0.975quant              mode
Min.    :1.61    Min.    :1.63    Min.    :1.61
1st Qu.:2.83    1st Qu.:2.86    1st Qu.:2.83
Median :3.05    Median :3.08    Median :3.05
Mean    :3.04    Mean    :3.06    Mean    :3.04
3rd Qu.:3.22    3rd Qu.:3.24    3rd Qu.:3.22
Max.    :3.91    Max.    :3.93    Max.    :3.91
```

We can create variables with the posterior mean (PM) and lower (LL) and upper (UL) limits of 95% credible intervals.

```
# Posterior mean and 95% CI
map$PM <- res$summary.fitted.values[, "mean"]
map$LL <- res$summary.fitted.values[, "0.025quant"]
map$UL <- res$summary.fitted.values[, "0.975quant"]
```

Then, we create maps of these variables with **mapview** specifying a common legend for the three maps, and using a popup table with the name, the logarithm of the housing prices, the covariates, and the posterior mean and 95% credible intervals. We use the sync() function of **leafsync** to plot synchronized maps (Figure 9.3).

```
# common legend
at <- seq(min(c(map$PM, map$LL, map$UL)),
          max(c(map$PM, map$LL, map$UL)),
          length.out = 8)

# popup table
popuptable <- leafpop::popupTable(dplyr::mutate_if(map,
                                  is.numeric, round, digits = 2),
zcol = c("TOWN", "vble", "CRIM", "RM", "PM", "LL", "UL"),
row.numbers = FALSE, feature.id = FALSE)

m1 <- mapview(map, zcol = "PM", map.types = "CartoDB.Positron",
              at = at, popup = popuptable)
m2 <- mapview(map, zcol = "LL", map.types = "CartoDB.Positron",
              at = at, popup = popuptable)
m3 <- mapview(map, zcol = "UL", map.types = "CartoDB.Positron",
              at = at, popup = popuptable)


library(leafsync)
m <- leafsync::sync(m1, m2, m3, ncol = 3)
m
```

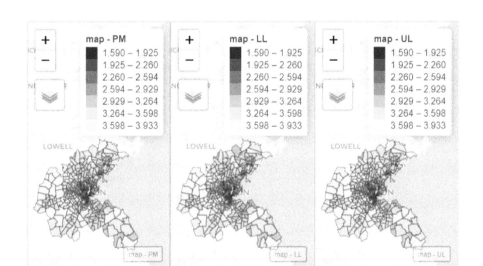

FIGURE 9.3: Posterior mean of the logarithm of the housing prices (left), together with lower (center) and upper (right) limits of 95% credible intervals.

We now obtain estimates of housing prices in their original scale by transforming the estimates of the logarithm of housing prices. First, we use the inla.tmarginal() function to obtain the marginals of the prices as exp(log(price)). Then, we use inla.zmarginal() to obtain the summaries of the marginals. Finally, we create variables PMoriginal, LLoriginal and ULoriginal with the posterior mean and lower and upper limits of 95% credible intervals of the posterior distribution of housing prices.

```
# Transformation of the marginal of
# the first area with inla.tmarginal()
# inla.tmarginal(function(x) exp(x),
#                res$marginals.fitted.values[[1]])

# Transformation marginals with inla.tmarginal()
marginals <- lapply(res$marginals.fitted.values,
FUN = function(marg){inla.tmarginal(function(x) exp(x), marg)})

# Obtain summaries of the marginals with inla.zmarginal()
marginals_summaries <- lapply(marginals,
FUN = function(marg){inla.zmarginal(marg)})

# Posterior mean and 95% CI
map$PMoriginal <- sapply(marginals_summaries, '[[', "mean")
map$LLoriginal <- sapply(marginals_summaries, '[[', "quant0.025")
map$ULoriginal <- sapply(marginals_summaries, '[[', "quant0.975")
```

Figure 9.4 shows maps with the estimated prices and lower and upper 95% credible intervals. The inspection of these maps allows us to understand the spatial pattern of housing prices in Boston, as well as the uncertainty in the estimates.

```
# common legend
at <- seq(min(c(map$PMoriginal, map$LLoriginal, map$ULoriginal)),
          max(c(map$PMoriginal, map$LLoriginal, map$ULoriginal)),
          length.out = 8)

# popup table
popuptable <- leafpop::popupTable(dplyr::mutate_if(map,
                                  is.numeric, round, digits = 2),
zcol = c("TOWN", "vble", "CRIM", "RM", "PM", "LL", "UL"),
row.numbers = FALSE, feature.id = FALSE)

m1 <- mapview(map, zcol = "PMoriginal",
              map.types = "CartoDB.Positron",
```

```
                        at = at, popup = popuptable)
m2 <- mapview(map, zcol = "LLoriginal",
                   map.types = "CartoDB.Positron",
                   at = at, popup = popuptable)
m3 <- mapview(map, zcol = "ULoriginal",
                   map.types = "CartoDB.Positron",
                   at = at, popup = popuptable)


m <- leafsync::sync(m1, m2, m3, ncol = 3)
m
```

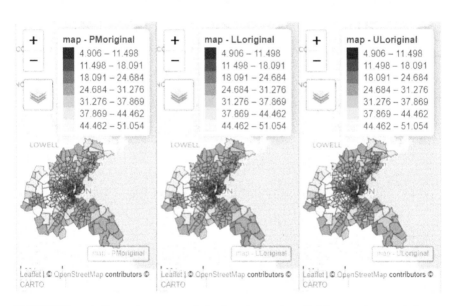

FIGURE 9.4: Posterior mean of the housing prices (left), together with lower (center) and upper (right) limits of 95% credible intervals.

10

Disease risk modeling

Areal data is common in disease mapping applications where often, for confidentiality reasons, individual incidence or mortality information is only available as the number of disease cases aggregated in areas. Disease data can be used to construct atlases that show the geographic distribution of aggregated outcomes to understand spatial patterns, identify high-risk areas, and reveal inequalities (Moraga, 2021a). For example, Bayesian spatial models have been used to understand geographic patterns and risk factors of childhood overweight and obesity prevalence in Costa Rica (Gómez et al., 2023), and mosquito-borne diseases in Brazil (Pavani et al., 2023). Spatial methods can also be extended to analyze areal data that are both spatially and temporally referenced. For example, Moraga and Kulldorff (2016) proposes a scan statistics method to detect spatial variations of temporal trends. Moraga and Ozonoff (2013) develop a spatio-temporal model-based imputation approach to produce more accurate estimates of all-cause and pneumonia and influenza mortality burden in the USA.

Disease risk can be estimated using Standardized Mortality Ratios (SMR) computed as the ratios of the observed to the expected number of mortality cases. Standardized Incidence Ratios (SIR) can be used when cases represent incidence data. However, these values may be extreme and unreliable in small populations and/or when dealing with rare diseases. Bayesian hierarchical models can instead be used to obtain smoothed relative risks by incorporating risk factors and borrowing information from neighboring areas (Moraga, 2019).

In this chapter, we demonstrate how to specify, fit, and interpret a Bayesian spatial model to estimate the risk of lung cancer and assess its relationship with smoking in Pennsylvania, USA, in 2002. Specifically, we show how to calculate the expected number of counts and SMR values, and how to obtain disease risk estimates and quantify risk factors using **R-INLA** (Rue et al., 2022). We also show how to make interactive maps of disease risk estimates using **mapview** (Appelhans et al., 2022). Moraga (2019) provides additional examples on how to fit both spatial and spatio-temporal models using **R-INLA** to understand geographic and temporal patterns of diseases, and assess their relationships with potential risk factors.

10.1   Spatial disease risk models

Bayesian hierarchical models enable to obtain smoothed disease relative risks by
including covariates and random effects to borrow information from neighboring
areas. Spatial disease risk models are commonly specified using a Poisson
distribution for the observed number of cases (Y_i) with mean equal to the
expected number of cases (E_i) times the relative risk (θ_i) corresponding to
area i, $i = 1, \ldots, n$,

$$Y_i \sim Poisson(E_i \times \theta_i), \; i = 1, \ldots, n,$$

$$\log(\theta_i) = \boldsymbol{z_i}\boldsymbol{\beta} + u_i + v_i.$$

Here, the logarithm of θ_i is expressed as a sum of fixed effects to quantify
the effects of the covariates on the disease risk, and random effects that
represent residual variation that is not explained by the available covariates.
The fixed effects $\boldsymbol{z_i}\boldsymbol{\beta}$ are expressed using a vector of intercept and p covariates
corresponding to area i, $\boldsymbol{z_i} = (1, z_{i1}, \ldots, z_{ip})$, and a coefficient vector $\boldsymbol{\beta} =
(\beta_0, \ldots, \beta_p)'$.

Spatial random effects u_i that smooth data according to a neighborhood
structure are included to acknowledge that data may be spatially correlated,
and relative risks in neighboring areas may be more similar than relative risks
in areas that are further away (Moraga and Lawson, 2012; Lawson et al.,
2015). Unstructured exchangeable components v_i are also included to model
uncorrelated noise.

The relative risk θ_i quantifies whether an area i has higher $(\theta_i > 1)$ or lower
$(\theta_i < 1)$ risk than the average risk in the standard population (e.g., the whole
population of the study region). For example, $\theta_i = 2$ indicates the risk of area
i is two times the average risk in the standard population.

10.2   Modeling of lung cancer risk in Pennsylvania

10.2.1   Data and map

Data with the number of lung cancer cases, population, as well as the smoking
proportions in the counties of Pennsylvania, USA, in 2002 can be obtained
from the **SpatialEpi** package (Kim et al., 2021). We load the **SpatialEpi**

package and attach the **pennLC** data. Reading the **pennLC** information with
?pennLC, we see **pennLC** is a list object with several elements.

```
library(SpatialEpi)
data(pennLC)
class(pennLC)
```

```
[1] "list"
```

```
names(pennLC)
```

```
[1] "geo"              "data"
[3] "smoking"          "spatial.polygon"
```

pennLC$data contains the number of lung cancer cases and the population
at county level, stratified in race (white and non-white), gender (female and
male) and age (under 40, 40-59, 60-69, and 70+).

```
head(pennLC$data)
```

```
  county cases population race gender      age
1  adams     0       1492    o      f Under.40
2  adams     0        365    o      f    40.59
3  adams     1         68    o      f    60.69
4  adams     0         73    o      f      70+
5  adams     0      23351    w      f Under.40
6  adams     5      12136    w      f    40.59
```

pennLC$smoking contains the proportion of smokers in each county.

```
head(pennLC$smoking)
```

```
     county smoking
1     adams   0.234
2 allegheny   0.245
3 armstrong   0.250
4    beaver   0.276
5   bedford   0.228
6     berks   0.249
```

pennLC$spatial.polygon is a **SpatialPolygons** object (sp object) with the
map of Pennsylvania counties. We create a map of class **sf** by converting
the **SpatialPolygons** object to a **sf** object with the st_as_sf() function of
sf (Pebesma, 2022a). We also add a column with the county names which
corresponds to the polygons ID slot values of **pennLC$spatial.polygon**.

```
library(sf)
map <- st_as_sf(pennLC$spatial.polygon)
countynames <- sapply(slot(pennLC$spatial.polygon, "polygons"),
                      function(x){slot(x, "ID")})
map$county <- countynames
head(map)
```

```
Simple feature collection with 6 features and 1 field
Geometry type: POLYGON
Dimension:     XY
Bounding box:  xmin: -80.52 ymin: 39.73 xmax: -75.53 ymax: 41.14
Geodetic CRS:  +proj=longlat
                        geometry     county
1 POLYGON ((-77.45 39.97, -77...      adams
2 POLYGON ((-80.15 40.67, -79... allegheny
3 POLYGON ((-79.21 40.91, -79... armstrong
4 POLYGON ((-80.16 40.85, -80...     beaver
5 POLYGON ((-78.38 39.73, -78...    bedford
6 POLYGON ((-75.53 40.45, -75...      berks
```

Now, we create a data frame called **d** with columns containing, for each of the counties, the county id (**county**), observed number of cases (**Y**), expected number of cases (**E**), smoking proportion (**smoking**), and SMR (**SMR**).

10.2.2 Observed cases

pennLC$data contains the cases in each county stratified by race, gender and age. We obtain the number of cases in each county, **Y**, by using the **group_by()** function of **dplyr** (Wickham et al., 2022b) to aggregate the rows of **pennLC$data** by county, and add up the observed number of cases.

```
library(dplyr)
d <- group_by(pennLC$data, county) %>% summarize(Y = sum(cases))
head(d)
```

```
# A tibble: 6 x 2
  county         Y
  <fct>      <int>
1 adams         55
2 allegheny   1275
3 armstrong     49
4 beaver       172
5 bedford       37
6 berks        308
```

10.2.3   Expected cases

The expected number of cases of a given area i represents the total number of cases that one would expect if the population in area i behaves in the same way as the standard population behaves (Moraga, 2018a). Typically, the standard population is considered as the whole population of all areas in the study region, and it is stratified in a number of groups. In this case, the standard population is considered as the whole population of Pennsylvania putting all counties together, and it is stratified in race, gender, and age groups.

The expected number of cases E_i in each county i can be calculated using indirect standardization as

$$E_i = \sum_{j=1}^{m} r_j^{(s)} n_j^{(i)},$$

where

$$r_j^{(s)} = \frac{\text{number of cases in group } j \text{ in standard population}}{\text{population in group } j \text{ in standard population}}$$

is the rate in group j in the standard population (Pennsylvania), and $n_j^{(i)}$ is the population in group j of county i. The number of expected counts can be easily obtained using the **expected()** function of **SpatialEpi** passing the following arguments:

- **population**: a vector of population counts for each group in each area,
- **cases**: a vector with the number of cases for each group in each area,
- **n.strata**: number of groups considered.

In **expected()**, vectors **population** and **cases** have to be sorted by area first and then, within each area, the counts for all groups need to be listed in the same order. The vectors need to include all groups so elements for groups with no cases need to be included as 0. Here, we use **order()** to sort the data by county, race, gender, and finally age.

```
pennLC$data <- pennLC$data[order(pennLC$data$county,
pennLC$data$race, pennLC$data$gender, pennLC$data$age), ]
```

Then, we obtain the expected counts **E** in each county using the **expected()** function passing the population **pennLC$data$population** and cases **pennLC$data$cases**. The number of groups is set to 16 since for each county there are 2 races, 2 genders, and 4 age groups ($2 \times 2 \times 4 = 16$).

```
E <- expected(population = pennLC$data$population,
              cases = pennLC$data$cases, n.strata = 16)
```

Finally, the vector with the expected counts E is included in the data frame d
that contains the counties ids (county) and the observed counts (Y).

```
d$E <- E
head(d)

# A tibble: 6 x 3
  county         Y      E
  <fct>      <int>  <dbl>
1 adams         55   69.6
2 allegheny   1275 1182.
3 armstrong     49   67.6
4 beaver       172  173.
5 bedford       37   44.2
6 berks        308  301.
```

10.2.4   Smokers proportions

In the spatial model, we will include the proportion of smokers as a covariate
to be able to quantify the effect of this factor. This variable is given by
pennLC$smoking, and we can add it to the data frame d that contains the rest
of the data as follows:

```
d <- dplyr::left_join(d, pennLC$smoking, by = "county")
```

10.2.5   Standardized Mortality Ratios

Let Y_i and E_i be the observed and expected number of cases, respectively, in
area $i, i = 1, \ldots, n$. The SMR in area i is defined as the ratio of the observed
to the expected number of cases,

$$\text{SMR}_i = \frac{Y_i}{E_i}, i = 1, \ldots, n.$$

If $\text{SMR}_i > 1$, this indicates there are more cases observed than expected which
corresponds to a high risk area. Similarly, a $\text{SMR}_i < 1$ indicates there are
fewer cases observed than expected. This corresponds to a low risk area. In
our example, SMRs are easily computed as the ratios of the observed to the
expected counts as follows:

```
d$SMR <- d$Y/d$E
```

The final data frame d contains the observed and expected disease counts, the
smokers proportions, and the SMR for each of the counties.

```
head(d)
```

```
# A tibble: 6 x 5
  county        Y        E smoking   SMR
  <fct>     <int>    <dbl>    <dbl> <dbl>
1 adams        55     69.6    0.234 0.790
2 allegheny  1275   1182.     0.245 1.08
3 armstrong    49     67.6    0.25  0.725
4 beaver      172    173.     0.276 0.997
5 bedford      37     44.2    0.228 0.837
6 berks       308    301.     0.249 1.02
```

10.2.6 Mapping SMR

To be able to make maps of the variables in **d**, we join the map and the data using the `left_join()` function of **dplyr** joining by the county id (by = "county"). Note that we could specify two different column names (by = c(name1, name2)) in case the column names were different in each of the objects to be joined.

```
map <- dplyr::left_join(map, d, by = "county")
```

We create an interactive choropleth map with the SMR values using the **mapview** package specifying the column name to plot in `zcol`. This map can be customized in several ways. For example, we can change the color border of the polygons with `color`, the opacity of the polygons with `alpha.regions`, and the legend title with `layer.name`. We can also add a color palette with `col.regions` and change the default base map with `map.types` using some of the options provided at https://leaflet-extras.github.io/leaflet-providers/preview/.

Figure 10.1 shows the map of SMR values created using opacity equal to a value less than 1 to be able to see the background map, and colors from the palette "YlOrRd" using the `brewer.pal()` function of the **RColorBrewer** package (Neuwirth, 2022).

```
library(mapview)
library(RColorBrewer)
pal <- colorRampPalette(brewer.pal(9, "YlOrRd"))
mapview(map, zcol = "SMR", color = "gray", alpha.regions = 0.8,
        layer.name = "SMR", col.regions = pal,
        map.types = "CartoDB.Positron")
```

FIGURE 10.1: SMRs of the counties of Pennsylvania, USA.

We can also highlight the counties when the mouse hovers over using `leaflet::highlightOptions()`, and setting `mapviewOptions(fgb = FALSE)`. In addition, we can customize the popups with tables showing information for each of the counties. This information that can be inspected by clicking each of the map polygons.

For example, here we use popups to show the values of the observed and expected counts, SMRs, and smoking proportions. To do that, we use the `popupTable()` function from **leafpop** (Appelhans and Detsch, 2021) which creates HTML strings to be used as popup tables in **mapview** (Appelhans et al., 2022) and **leaflet** (Cheng et al., 2022a). We create the popup table by passing the spatial object `map` with the numeric values rounded to two digits, the vector `zcol` indicating the columns to be included in the table, and setting `row.numbers = FALSE` and `feature.id = FALSE` to hide row numbers and feature ids, respectively.

```
library(mapview)
library(RColorBrewer)
library(leafpop)
```

```
pal <- colorRampPalette(brewer.pal(9, "YlOrRd"))
mapviewOptions(fgb = FALSE)

popuptable <- leafpop::popupTable(dplyr::mutate_if(map,
is.numeric, round, digits = 2),
zcol = c("county", "Y", "E", "smoking", "SMR"),
row.numbers = FALSE, feature.id = FALSE)

mapview(map, zcol = "SMR", color = "gray", col.regions = pal,
highlight = leaflet::highlightOptions(weight = 4),
popup = popuptable)
```

The map with the SMR values allows us to understand the spatial pattern of lung cancer risk across Pennsylvania, and identify areas that have SMR higher (or lower) than 1 indicating the observed cases are higher (or lower) than expected from the standard population. As we have seen, SMR values can be easily calculated as the ratio of observed to expected counts. However, these values may be extreme and unreliable for reporting in areas with small populations or rare diseases. To overcome these limitations, we use Bayesian hierarchical models that enable to incorporate covariates known to affect disease risk and borrow information from neighboring areas to obtain smoothed relative risks. Below, we show how to specify, fit, and interpret a spatial model to estimate the risk of lung cancer.

10.2.7 Model

Let Y_i and E_i be the observed and expected number of disease cases, respectively, and let θ_i be the relative risk for county $i = 1, \ldots, n$. The model is specified as follows:

$$Y_i | \theta_i \sim Poisson(E_i \times \theta_i), \ i = 1, \ldots, n,$$

$$\log(\theta_i) = \beta_0 + \beta_1 \times smoking_i + u_i + v_i.$$

Here, β_0 is the intercept and β_1 is the coefficient of the covariate smokers proportion. u_i is a structured spatial effect modeled with an intrinsic conditionally autoregressive model (CAR), $u_i | \mathbf{u}_{-i} \sim N(\bar{u}_{\delta_i}, \frac{1}{\tau_u n_{\delta_i}})$. Finally, v_i is an unstructured effect, $v_i \sim N(0, 1/\tau_v)$.

10.2.8   Neighborhood matrix

The spatial random effect u_i needs the specification of the neighborhood matrix. Here, we assume two areas are neighbors if they share a common boundary. We can obtain the neighbourhood list using the poly2nb() function of the **spdep** package (Bivand, 2022). Then, we use the nb2INLA() and inla.read.graph() functions to create an object g with the neighborhood matrix in the format required by **R-INLA**.

```
library(spdep)
library(INLA)
nb <- poly2nb(map)
nb2INLA("map.adj", nb)
g <- inla.read.graph(filename = "map.adj")
```

10.2.9   Model formula and inla() call

The model formula is specified by writing the outcome variable, the ~ symbol, and the covariates and random effects. An intercept is included in the model by default. In the formula, random effects are set using the f() function with arguments equal to indices vectors of the variables, and the model name. The indices for the random effects are given by indices vectors re_u and re_v created for the random effects u_i and v_i, respectively. These vectors are equal to $1, \ldots, n$, where n is the number of counties. Here, number of counties $n=67$ can be obtained with the number of rows in the data (nrow(map)). For the spatial random effect u_i, we use model = "besag" with neighborhood matrix given by g. For the unstructured effect v_i we choose model = "iid".

```
map$re_u <- 1:nrow(map)
map$re_v <- 1:nrow(map)
```

```
formula <- Y ~ smoking +
  f(re_u, model = "besag", graph = g, scale.model = TRUE) +
  f(re_v, model = "iid")
```

Then, we fit the model using the inla() function with the default priors in **R-INLA**. We specify the formula, family, data, and the expected counts, and set control.predictor = list(compute = TRUE) and control.compute = list(return.marginals.predictor = TRUE) to compute and return the posterior means of the predictors.

```
res <- inla(formula, family = "poisson", data = map, E = E,
control.predictor = list(compute = TRUE),
control.compute = list(return.marginals.predictor = TRUE))
```

10.2.10   Results

The inla() function returns an object res with the fit of the model that can be inspected with using summary(res). Objects res$summary.fixed, res$summary.random, and res$summary.hyperpar contain, respectively, summaries of the fixed effects, random effects, and the hyperparameters.

res$summary.fixed

```
              mean       sd 0.025quant 0.5quant
(Intercept) -0.3235 0.1498   -0.61925  -0.3233
smoking      1.1546 0.6226   -0.07569   1.1560
            0.975quant     mode        kld
(Intercept)  -0.02877 -0.3234 3.534e-08
smoking       2.37845  1.1563 3.545e-08
```

We see the intercept $\hat{\beta}_0 = -0.323$ with a 95% credible interval equal to $(-0.619, -0.029)$, and the coefficient of smoking is $\hat{\beta}_1 = 1.155$ with a 95% credible interval equal to $(-0.076, 2.378)$ This indicates a non-significant effect of smoking.

res$summary.fitted.values contains the posterior mean and quantiles of the relative risk of each of the counties, θ_i, $i = 1, \ldots, n$. We add to map the disease relative risk estimates which are given by the posterior mean (column mean of res$summary.fitted.values). We also add to map the 2.5 and 97.5 percentiles of the posterior distribution which are given by columns 0.025quant and 0.975quant of res$summary.fitted.values. These percentiles represent the lower and upper limits of 95% credible intervals of the risks representing the uncertainty of the risks estimated.

res$summary.fitted.values[1:3,]

```
                         mean       sd 0.025quant 0.5quant
fitted.Predictor.01 0.8781 0.05808     0.7648   0.8778
fitted.Predictor.02 1.0597 0.02750     1.0072   1.0592
fitted.Predictor.03 0.9646 0.05089     0.8604   0.9657
                     0.975quant     mode
fitted.Predictor.01    0.9936 0.8778
fitted.Predictor.02    1.1150 1.0582
fitted.Predictor.03    1.0622 0.9681
```

```
# relative risk
map$RR <- res$summary.fitted.values[, "mean"]
# lower and upper limits 95% CI
map$LL <- res$summary.fitted.values[, "0.025quant"]
map$UL <- res$summary.fitted.values[, "0.975quant"]
```

10.2.11   Mapping disease risk

Figure 10.2 shows the estimated relative risks (RRs) in an interactive map
using **mapview**. In the map, we add popups showing information on the
observed and expected counts, SMRs, smokers proportions, RRs, and limits of
95% credible intervals. We observe counties with greater RR are located in the
west and south-east of Pennsylvania, and counties with lower RR are located
in the center. The 95% credible intervals indicate the uncertainty in the RRs.

```
library(mapview)
library(RColorBrewer)
library(leafpop)
pal <- colorRampPalette(brewer.pal(9, "YlOrRd"))
mapviewOptions(fgb = FALSE)
mapview(map, zcol = "RR", color = "gray", col.regions = pal,
highlight = leaflet::highlightOptions(weight = 4),
popup = leafpop::popupTable(dplyr::mutate_if(map, is.numeric,
                                          round, digits = 2),
zcol = c("county", "Y", "E", "smoking", "SMR", "RR", "LL", "UL"),
row.numbers = FALSE, feature.id = FALSE))
```

FIGURE 10.2: Relative risks of the counties of Pennsylvania, USA.

10.2.12 Comparing SMR and RR maps

We compare the maps of SMRs and RRs using side-by-side synchronized maps with the same scale created with the **leafsync** package (Appelhans and Russell, 2019). We see the RR values are shrunk towards 1 compared to the SMR values (Figure 10.3).

```
at <- seq(min(map$SMR), max(map$SMR), length.out = 8)
m1 <- mapview(map, zcol = "SMR", color = "gray",
              col.regions = pal, at = at)
m2 <- mapview(map, zcol = "RR", color = "gray",
              col.regions = pal, at = at)
leafsync::sync(m1, m2)
```

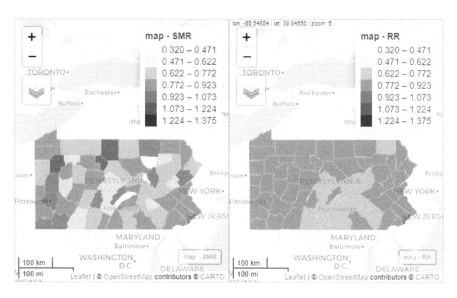

FIGURE 10.3: SMRs (left) and RRs (right) of the counties of Pennsylvania, USA.

10.2.13 Exceedance probabilities

In addition to the relative risks, we can also calculate exceedance probabilities that allow us to assess unusual elevation of disease risk. Exceedance probabilities are defined as the probabilities of relative risk being greater than a given threshold value c. For example, we can calculate the probability that the relative risk of the 51st county (Philadelphia) exceeds $c = 1.2$ as $P(\theta_{51} > c) = 1 - P(\theta_{51} \leq c)$. We can calculate this exceedance probability using the `inla.pmarginal()` function passing the marginal distribution of θ_{51} and the threshold value $c = 1.2$ as follows:

```
c <- 1.2
marg <- res$marginals.fitted.values[[51]]
1 - inla.pmarginal(q = c, marginal = marg)
```

```
[1] 0.05616
```

We can plot the posterior distribution of θ_{51} by first calculating a smoothing of the marginal distribution with `inla.smarginal()`, and then using **ggplot2** (Figure 10.4). $P(\theta_{51} > c)$ is the area under the curve to the right of the threshold value c.

```
library(ggplot2)
marginal <- inla.smarginal(res$marginals.fitted.values[[51]])
marginal <- data.frame(marginal)
ggplot(marginal, aes(x = x, y = y)) + geom_line() +
  labs(x = expression(theta[51]), y = "Density") +
  geom_vline(xintercept = 1.2, col = "black") +
  theme_bw(base_size = 20)
```

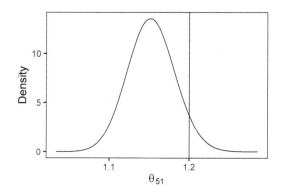

FIGURE 10.4: Posterior distribution of the relative risk of area 51 exceeds the threshold value 1.2. Vertical line indicates the threshold value.

To calculate the exceedance probabilities for all counties, we can use the `sapply()` function as follows:

```
c <- 1.2
map$exc <- sapply(res$marginals.fitted.values,
FUN = function(marg){1 - inla.pmarginal(q = c, marginal = marg)})
```

Figure 10.5 shows a map with the exceedance probabilities created with **mapview**. The map provides evidence of excess risk within individual areas.

In areas with probabilities close to 1, it is very likely that the relative risk exceeds the threshold value c, and areas with probabilities close to 0 correspond to areas where it is very unlikely that the relative risk exceeds c. Areas with probabilities around 0.5 have the highest uncertainty, and they correspond to areas where the relative risk is below or above c with equal probability. In the map depicting the exceedance probabilities, we observe all probabilities are close to 0 and it is very unlikely the relative risk exceeds the threshold value c in any of the counties.

```
pal <- colorRampPalette(brewer.pal(9, "YlOrRd"))
mapview(map, zcol = "exc", color = "gray", col.regions = pal,
        map.types = "CartoDB.Positron")
```

FIGURE 10.5: Probabilities that the relative risks of counties exceed 1.2.

11

Areal data issues

Spatial analyses of aggregated data may be subject to the Misaligned Data Problem (MIDP) which refers to a situation where the spatial data being analyzed are at a different scale than the one at which they were originally collected (Banerjee et al., 2004). For example, individual observations or small areas data may be aggregated to larger areas due to several reasons such as confidentiality or to match the scale of other data sources. The aggregation of the data can lead to a loss of spatial information that can hide spatial patterns or relationships that exist at the finer scale, potentially leading to erroneous conclusions or misinterpretation of the results.

The Modifiable Areal Unit Problem (MAUP) (Openshaw, 1984) refers to the issue of how the results of spatial analyses may change if one aggregates the same underlying data to a different level of spatial aggregation. The MAUP consists of two interrelated effects, namely, the scale and zoning effects. The MAUP's scale effect occurs when the results of an analysis change because the geographic units used for analysis are aggregated or disaggregated. For example, if a study examines crime rates across different neighborhoods, the patterns observed may differ depending on whether the analysis is conducted at the level of city blocks, census tracts, or larger administrative zones. The zoning effect of the MAUP arises when the results of an analysis are impacted by the arbitrary creation of geographic units. For example, when examining income levels across various districts, the specific boundaries assigned to each district can influence the results. Thus, different configurations of boundaries can produce different spatial patterns, leading to variations in the outcomes of the analysis.

Ecological studies, also known as population-level studies, investigate the relationships between exposure factors or interventions and health outcomes at the group or population level (Robinson, 1950). Instead of focusing on individual-level data, ecological studies analyze aggregated data for groups or populations, such as cities, regions, or countries. Ecological studies are useful when individual-level data is not available or difficult to collect. However, ecological studies face the ecological fallacy, where associations observed at the group level may not hold true for individuals within those groups. The resulting bias, known as ecological bias, can be viewed as a special case of the MAUP, as it encompasses two effects similar to the aggregation and zoning effects in the MAUP (Gotway and Young, 2002). These effects are the aggregation bias, which arises from grouping individuals together, and the specification bias, resulting from the uneven distribution of confounding variables resulting from grouping.

Measurements of a spatial phenomenon can be obtained at various spatial resolutions and from diverse sources. For instance, air pollution measurements can be gathered from monitoring stations located at specific locations, as well as through satellite-derived measurements that provide aggregated information in areas. The integration of these data can lead to more accurate air pollution predictions at finer spatial resolutions than the ones obtained using just one type of data. Moraga et al. (2017) proposes a Bayesian melding model to combine spatially misaligned data that assumes a common spatially continuous Gaussian random field underlying all observations, and uses the integrated nested Laplace approximation (INLA) and stochastic partial differential equation (SPDE) approaches for fast inference. Zhong and Moraga (2023) compare the Bayesian melding model with a Bayesian downscaler approach that integrates point- and area-level data by considering a model with spatially varying coefficients that has point data as response and areal data as covariates. They also use air pollution data to show how the melding model can be used to disaggregate areal data and produce spatially continuous predictions, as well as predictions at certain spatial resolutions that are policy relevant improving decision-making.

Part III

Geostatistical data

12

Geostatistical data

Geostatistical data provide information of a spatially continuous phenomenon that has been measured at particular sites. This type of data may represent, for example, air pollution levels taken at a set of monitoring stations or disease prevalence survey data at a collection of sites. Let $Z(s_1), \ldots, Z(s_n)$ be observations of a spatial variable Z at locations $s_1, \ldots, s_n$. In many situations, geostatistical data may be assumed to be a partial realization of a random process

$$\{Z(s) : s \in D \subset \mathbb{R}^2\},$$

where D is a fixed subset of $\mathbb{R}^2$ and the spatial index s varies continuously throughout D. Often, for practical reasons, the process $Z(\cdot)$ can only be observed at a finite set of locations. Based upon this partial realization, we may seek to infer the characteristics of the spatial process that gives rise to the data observed such as the mean and variability of the process. Then, we can use this information to predict the process at unsampled locations and construct a spatially continuous surface of the variable of study.

Kriging (Matheron, 1963) and model-based geostatistics (Diggle et al., 1998) are widely used approaches for spatial interpolation. Both approaches model the spatial distribution of the data to obtain predictions and their associated uncertainty at unsampled locations. Simpler spatial interpolation methods, such as the inverse distance weighted method, obtain predictions based on the spatial arrangement of the data. In the following chapters, we give an overview of Gaussian random fields. Then, we show how to employ spatial interpolation methods to obtain predictions using geostatistical data. We also show how to evaluate the predictive performance of the methods using a number of error measures and cross-validation.

12.1   Gaussian random fields

A Gaussian random field (GRF) $\{Z(s) : s \in D \subset \mathbb{R}^2\}$ is a collection of random variables where observations occur in a continuous domain, and where every finite collection of random variables has a multivariate normal distribution. A random process $Z(\cdot)$ is said to be strictly stationary if it is invariant to shifts. That is, if for any set of locations s_i, $i = 1, \ldots, n$, and any $h \in \mathbb{R}^2$ the distribution of $\{Z(s_1), \ldots, Z(s_n)\}$ is the same as that of $\{Z(s_1 + h), \ldots, Z(s_n + h)\}$. A less restrictive condition is given by the second-order stationarity (or weakly stationarity). Under this condition, the process has constant mean,

$$E[Z(s)] = \mu, \forall s \in D,$$

and the covariances depend only on the differences between locations,

$$Cov(Z(s), Z(s + h)) = C(h), \forall s \in D, \forall h \in \mathbb{R}^2.$$

In addition, if the covariances are functions only of the distances between locations $h = \|h\|$ and not of the directions, the process is called isotropic. If not, it is anisotropic. A process is said to be intrinsically stationary if, in addition to the constant mean assumption, it satisfies

$$Var[Z(s_i) - Z(s_j)] = 2\gamma(s_i - s_j), \forall s_i, s_j.$$

12.2 Covariance functions of Gaussian random fields

Given a stationary and isotropic Gaussian random process $Z(\cdot)$, the covariance function between a pair of variables separated by a distance h can be expressed as

$$C(h) = \sigma^2 \text{Corr}(h),$$

where σ^2 denotes the variance of the spatial field, and $\{\text{Corr}(h) : h \in \mathbb{R}\}$ is a positive definite correlation function.

The Matérn family represents a very flexible class of correlation functions that appears naturally in many scientific fields. This family can be specified as

$$\text{Corr}(h) = \frac{1}{2^{\nu-1}\Gamma(\nu)} \left(\frac{h}{\phi}\right)^\nu K_\nu \left(\frac{h}{\phi}\right),$$

where h denotes the distance between locations, and $K_\nu(\cdot)$ is the modified Bessel function of second kind and order $\nu > 0$. The parameter $\phi > 0$ controls how fast the correlation decays with distance and is related to the range, the distance at which the correlation between two points is approximately 0. The parameter ν determines the smoothness of the process. This parameter is difficult to estimate in applications, and it is usually fixed to reflect scientific knowledge about the smoothness of the process $Z(\cdot)$ (Diggle and Ribeiro Jr., 2007).

Special classes of the Matérn family include the exponential correlation function $\text{Corr}(h) = exp(-h/\phi)$ when $\nu = 0.5$, and the squared exponential or Gaussian correlation function $\text{Corr}(h) = exp(-(h/\phi)^2)$ when $\nu \to \infty$.

12.3 Simulating Gaussian random fields

Here, we show how to generate realizations of several Gaussian random fields using the **geoR** package (Ribeiro Jr et al., 2022). Other packages that can be used to simulate Gaussian random fields include **RandomFields** (Schlather et al., 2015) and **gstat** (Pebesma and Graeler, 2022).

The `cov.spatial()` function of **geoR** computes the value of the covariance function $C(h) = \sigma^2\text{Corr}(h)$ between a pair of variables located at points separated by distance h, where σ^2 is the variance parameter and $\text{Corr}(h)$ is a positive definite correlation function. The types of covariance functions implemented in **geoR** can be seen by typing `?cov.spatial`.

The `grf()` function of **geoR** can be used to simulate realizations of a Gaussian random field by specifying the following arguments:

- `n` is the number of points in the simulation. Here, we choose 1024 to simulate in a grid with 32×32 cells.
- `grid` can be set to `"reg"` to simulate in a regular grid or `"irreg"` to simulate in a number of coordinates.
- `xlims` and `ylims` specify the limits of the area in the x and y directions, respectively. Here, we generate the field in a squared region $[0,1] \times [0,1]$ by using the default values `xlims = c(0, 1)` and `ylims = c(0, 1)`.
- `cov.model` denotes the type of correlation function (e.g., `"matern"`). Possible types are detailed in the `cov.spatial()` function.
- `cov.pars` specifies the covariance parameters that indicate the variance σ^2 and the parameter related to the range ϕ.
- `kappa` is the smoothness parameter required by some correlation functions such as `"matern"`. By default, `kappa = 0.5`. In particular, `kappa` represents ν in the definition of the Matérn function above.

Here, we show the correlation functions and realizations of Gaussian random fields with Matérn covariance function,

$$C(h) = \frac{\sigma^2}{2^{\nu-1}\Gamma(\nu)} \left(\frac{h}{\phi}\right)^{\nu} K_{\nu}\left(\frac{h}{\phi}\right),$$

with $\sigma^2 = 1$ and $\phi = 0.01$, 0.2 and 1 to show different patterns of spatial variability. Figure 12.1 shows the covariance functions and realizations corresponding to each of the Gaussian random fields. We observe the correlation goes to 0 with distance, and it goes faster to 0 with smaller values of ϕ. The realizations of Gaussian random fields show less spatial autocorrelation for smaller values of ϕ.

```
covmodel <- "matern"
sigma2 <- 1

# Plot covariance function
curve(cov.spatial(x, cov.pars = c(sigma2, 1),
                  cov.model = covmodel), lty = 1,
      from = 0, to = 1, ylim = c(0, 1), main = "Matérn",
      xlab = "distance", ylab = "Covariance(distance)")
curve(cov.spatial(x, cov.pars = c(sigma2, 0.2),
                  cov.model = covmodel), lty = 2, add = TRUE)
curve(cov.spatial(x, cov.pars = c(sigma2, 0.01),
                  cov.model = covmodel), lty = 3, add = TRUE)

legend(cex = 1.5, "topright", lty = c(1, 1, 2, 3),
       col = c("white", "black", "black", "black"),
       lwd = 2, bty = "n", inset = .01,
       c(expression(paste(sigma^2, " = 1 ")),
         expression(paste(phi, " = 1")),
         expression(paste(phi, " = 0.2")),
         expression(paste(phi, " = 0.01")))))

# Simulate Gaussian random field in a regular grid 32 X 32
sim1 <- grf(1024, grid = "reg",
            cov.model = covmodel, cov.pars = c(sigma2, 1))
sim2 <- grf(1024, grid = "reg",
            cov.model = covmodel, cov.pars = c(sigma2, 0.2))
sim3 <- grf(1024, grid = "reg",
            cov.model = covmodel, cov.pars = c(sigma2, 0.01))

# Plot Gaussian random field
par(mfrow = c(1, 3), mar = c(2, 2, 2, 0.2))
image(sim1, main = expression(paste(phi, " = 1")), cex.main = 2,
      col = gray(seq(1, 0.1, l = 30)), xlab = "", ylab = "")
image(sim2, main = expression(paste(phi, " = 0.2")), cex.main = 2,
      col = gray(seq(1, 0.1, l = 30)), xlab = "", ylab = "")
image(sim3, main = expression(paste(phi, " = 0.01")), cex.main=2,
      col = gray(seq(1, 0.1, l = 30)), xlab = "", ylab = "")
```

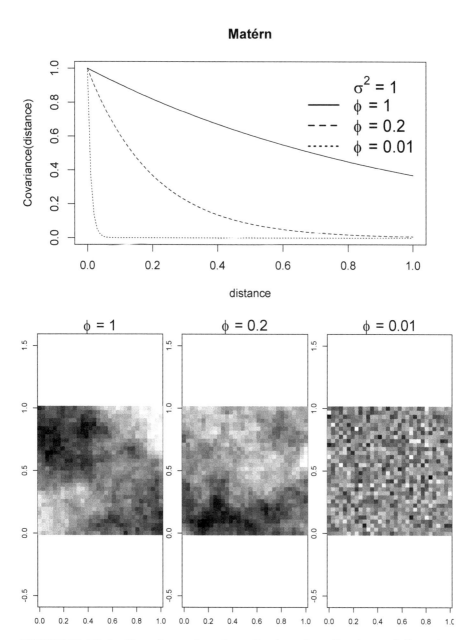

FIGURE 12.1: Covariance functions (top) and realizations of Gaussian random fields (bottom) for several parameter values.

12.4 Variogram

We can summarize the covariance structure of a spatial Gaussian random field with its variogram $2\gamma(\cdot)$ (or semivariogram $\gamma(\cdot)$). The variogram of a Gaussian random field $Z(\cdot)$ is defined as the function

$$Var[Z(\boldsymbol{s}_i) - Z(\boldsymbol{s}_j)] = 2\gamma(\boldsymbol{s}_i - \boldsymbol{s}_j).$$

Under the assumption of intrinsic stationarity, the constant-mean assumption implies

$$2\gamma(\boldsymbol{h}) = Var(Z(\boldsymbol{s} + \boldsymbol{h}) - Z(\boldsymbol{s})) = E[(Z(\boldsymbol{s} + \boldsymbol{h}) - Z(\boldsymbol{s}))^2],$$

and the semivariogram can be easily estimated with the empirical semivariogram as follows:

$$2\hat{\gamma}(\boldsymbol{h}) = \frac{1}{|N(\boldsymbol{h})|} \sum_{N(\boldsymbol{h})} (Z(\boldsymbol{s}_i) - Z(\boldsymbol{s}_j))^2,$$

where $|N(\boldsymbol{h})|$ denotes the number of distinct pairs in $N(\boldsymbol{h}) = \{(\boldsymbol{s}_i, \boldsymbol{s}_j) : \boldsymbol{s}_i - \boldsymbol{s}_j = \boldsymbol{h}, \ i, j = 1, \ldots, n\}$. Note that if the process is isotropic, the semivariogram is a function of the distance $h = ||\boldsymbol{h}||$.

The empirical semivariogram, when plotted against the separation distance, provides crucial insights into the continuity and spatial variability of the process. Figure 12.2 shows a plot of a typical semivariogram. Often, at shorter distances, the semivariogram tends to be small, indicating a higher similarity among observations in close proximity compared to those farther apart. As the separation distance increases, the semivariogram tends to increase as well, suggesting a decrease in similarity between observations with increasing distance.

Beyond a certain separation distance known as the range, the semivariogram levels off and reaches a nearly constant value referred to as the sill. This indicates that spatial dependence between observations decays with distance within the range, and beyond that range, observations become spatially uncorrelated, resulting in a near constant variance.

If there is a discontinuity or vertical jump at the origin of the plot, it signifies a nugget effect in the process. This effect is often attributed to measurement error but can also indicate a spatially discontinuous process.

The empirical semivariogram serves as a valuable exploratory tool for evaluating the presence of spatial correlation in data. Additionally, we can compare the empirical semivariogram to a Monte Carlo envelope constructed by computing

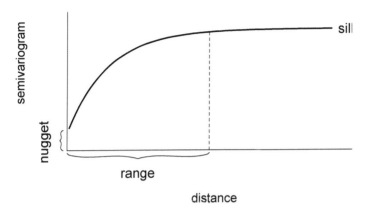

FIGURE 12.2: Typical semivariogram.

empirical semivariograms from random permutations of the data while keeping the locations fixed. If the empirical semivariogram shows an increasing trend with distance and lies outside the Monte Carlo envelope, it provides evidence of spatial correlation.

Example

Here, we show how to estimate the variogram function of geostatistical data using the variog() function of **geoR**. We use the **parana** data from **geoR** which contains the average rainfall over different years for the period May to June at 123 monitoring stations in Paraná state, Brazil. We use the st_as_sf() package to create a **sf** object with the rainfall data and create a map depicting the rainfall values with **ggplot2** (Figure 12.3).

```
library(geoR)
library(ggplot2)
library(sf)
d <- st_as_sf(data.frame(x = parana$coords[, 1],
                         y = parana$coords[, 2],
                         value = parana$data),
             coords = c("x", "y"))

ggplot(d) + geom_sf(aes(color = value), size = 2) +
scale_color_gradient(low = "blue", high = "orange") +
geom_path(data = data.frame(parana$border), aes(east, north)) +
theme_bw()
```

In variog(), the argument option specifies the type of variogram. Possible values of option are binned ("bin"), cloud ("cloud"), and kernel smoothed variogram ("smooth"). The variogram cloud of a set of geostatistical data

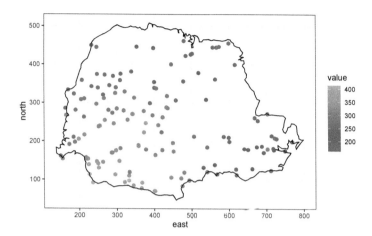

FIGURE 12.3: Rainfall values measured at 143 recording stations in Paraná state, Brazil.

is a scatterplot of the points (h_{ij}, v_{ij}) defined as $h_{ij} = ||\boldsymbol{s}_i - \boldsymbol{s}_j||$ and $v_{ij} = \frac{1}{2}(Z(\boldsymbol{s}_i) - Z(\boldsymbol{s}_j))^2$.

When the underlying process has a spatially varying mean $\mu(\boldsymbol{s})$, we compute v_{ij} using the residuals $(Z(\boldsymbol{s}_i) - \hat{\mu}(\boldsymbol{s}_i))$ instead of the data $Z(\boldsymbol{s}_i)$. Argument `trend` of `variog()` denotes the trend fitted using ordinary least squares so variograms are computed using the residuals. By default, `trend = "cte"` so the mean is assumed constant over the region.

Figure 12.4 shows the empirical variogram corresponding to the rainfall data in Paraná state. It also shows the empirical variogram obtained by averaging values v_{ij}'s for which $|h - h_{ij}| < u/2$, where u is a chosen bandwidth.

```
plot(variog(coords = st_coordinates(d), data = d$value,
            option = "cloud", max.dist = 400))

plot(variog(parana, max.dist = 400))
```

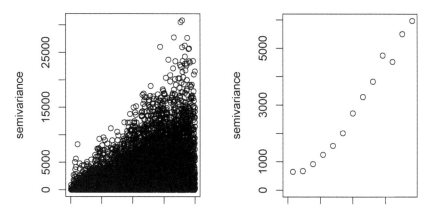

FIGURE 12.4: Empirical variogram values (left) and averaged empirical variogram values (right) corresponding to the rainfall data in Paraná state, Brazil.

The covariance parameters can be estimated by fitting a parametric covariance function to the empirical variogram. We can obtain the estimates by eye without a formal criterion, or by using ordinary or weighted least squares methods as we will see in the next chapters.

13

Spatial interpolation methods

In this chapter, we describe several simple interpolation methods that allow us to predict values of a spatially continuous variable at locations that are not sampled. These methods include closest observation, inverse distance weighting (IDW), and nearest neighbors, and can be easily implemented using the **gstat** (Pebesma and Graeler, 2022) and **terra** (Hijmans, 2022) packages. We also describe an ensemble approach to compute predictions by combining the predictions of several methods.

13.1 Spatial prediction of property prices

13.1.1 Data

To illustrate the spatial interpolation methods, we use the `properties` data of the **spData** package (Bivand et al., 2022) which contains the price of apartments in Athens, Greece, in 2017. `properties` is a `sf` object that contains several columns including `price` with the apartments' price in Euros, `prpsqm` with the apartments' price per square meter, and `geometry` with the coordinates of the locations. We create a `sf` object called `d` with the contents of `properties`, and a new variable `vble` with the variable of interest which in this case is price per square meter (`prpsqm`).

```
library(spData)
library(sf)
library(terra)
library(tmap)
library(viridis)
d <- properties
d$vble <- d$prpsqm
```

The object `depmunic` contains the boundaries of the administrative divisions of Athens. We create a variable `map` denoting the study region with the union of these boundaries. This is the region where we will predict the variable of interest price per square meter.

```
map <- st_union(depmunic) %>% st_sf()
```

The tmap_mode() function of **tmap** can be used to create static
(tmap_mode("plot")) or interactive (tmap_mode("view")) maps. Figure 13.1
shows a static map with the locations and prices of the apartments.

```
tmap_mode("plot")
tm_shape(map) + tm_polygons(alpha = 0.3) + tm_shape(d) +
  tm_dots("vble", palette = "viridis")
```

13.1.2   Prediction locations

We wish to predict the prices of properties continuously in space in Athens. To
do that, we create a fine raster grid covering Athens and consider the centroids
of the raster cells as the prediction locations. We create the raster grid with the
rast() function of **terra** by specifying the region to be covered (map), and the
number of rows and columns of the grid (nrows = 100, ncols = 100). The
prediction locations are the centroids of the raster cells and can be obtained
with the xyFromCell() function of **terra**. Alternatively, we could create the
grid using the st_make_grid() function of **sf** by specifying the number of
grid cells in the horizontal and vertical directions or the cell size.

```
library(sf)
library(terra)
# raster grid covering map
grid <- terra::rast(map, nrows = 100, ncols = 100)
# coordinates of all cells
xy <- terra::xyFromCell(grid, 1:ncell(grid))
```

We create a **sf** object called coop with the prediction locations with
st_as_sf() passing the coordinates (as.data.frame(xy)), the name of the coordinates (coords = c("x", "y")), and the coordinate reference system which
is the same as the coordinate reference system of map (crs = st_crs(map)).
Then, we use st_filter() to keep the locations within the Athens map. Figure 13.1 shows the map with the prediction locations created using the qtm()
function of **tmap** for quick map plots.

```
coop <- st_as_sf(as.data.frame(xy), coords = c("x", "y"),
                 crs = st_crs(map))
coop <- st_filter(coop, map)

qtm(coop)
```

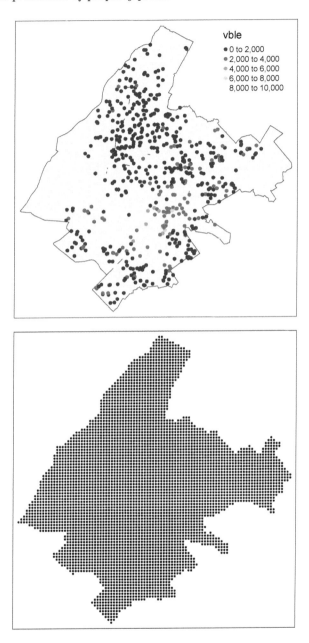

FIGURE 13.1: Top: Locations and prices per square meter of apartments in Athens. Bottom: Prediction locations.

13.1.3   Closest observation

We can obtain predictions at each of the prediction locations as the values of
the closest sampled locations. To do that, we can employ the Voronoi diagram
(also known as Dirichlet or Thiessen diagram). The Voronoi diagram is created
when a region with n points is partitioned into convex polygons such that
each polygon contains exactly one generating point, and every point in a given
polygon is closer to its generating point than to any other.

Given a set of points, we can create a Voronoi diagram with the `voronoi()`
function of **terra** specifying the points as an object of class `SpatVector` of
terra, and `map` to set the outer boundary of the Voronoi diagram. This returns
a Voronoi diagram for the set of points assuming constant values in each of
the polygons (Figure 13.2).

Then, we can use the functions `tm_shape()` and `tm_fill()` of **tmap** to plot
the values of the variable in each of the polygons indicating the name of the
variable `col = "vble"`, the palette `palette = "viridis"` and the level of
transparency `alpha = 0.6` (if the plot is interactive).

```
# Voronoi
v <- terra::voronoi(x = terra::vect(d), bnd = map)
plot(v)
points(vect(d), cex = 0.5)

# Prediction
v <- st_as_sf(v)
tm_shape(v) +
  tm_fill(col = "vble", palette = "viridis", alpha = 0.6)
```

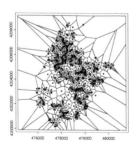 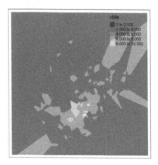

FIGURE 13.2: Left: Voronoi diagram corresponding to the observation
locations. Right: Predictions obtained using the closest observation criterion.

We can also extract the predicted values at prediction points `coop` by using the `st_intersection()` function of the **sf** package. Here, instead of creating a map with the predictions at the locations given by `coop`, we create a map with the predictions at a raster grid. To do that, we transfer the **sf** object with the predictions at locations `coop` to the raster grid `grid`, and create a plot with the raster values with **tmap** (Figure 13.3).

```
resp <- st_intersection(v, coop)
resp$pred <- resp$vble

pred <- terra::rasterize(resp, grid, field = "pred", fun = "mean")
tm_shape(pred) + tm_raster(alpha = 0.6, palette = "viridis")
```

FIGURE 13.3: Predictions obtained using closest observation (top-left), IDW (top-right), nearest neighbors (bottom-left), and ensemble approach (bottom-right).

13.1.4   IDW: Inverse Distance Weighting

In the IDW method, values at unsampled locations are estimated as the weighted average of values from the rest of locations with weights inversely proportional to the distance between the unsampled and the sampled locations. Specifically,

$$\hat{Z}(\boldsymbol{s}_0) = \frac{\sum_{i=1}^{n} Z(\boldsymbol{s}_i) \times (1/d_i^{\beta})}{\sum_{i=1}^{n}(1/d_i^{\beta})} = \sum_{i=1}^{n} Z(\boldsymbol{s}_i) \times w_i,$$

where $\hat{Z}(\boldsymbol{s}_0)$ is the predicted value at $\boldsymbol{s}_0$, n the number of sampled locations, $Z(\boldsymbol{s}_i)$ is the value at location $\boldsymbol{s}_i$, and d_i the distance between location $\boldsymbol{s}_i$ and location $\boldsymbol{s}_0$ where we want to predict. Here, weights are given by $w_i = \frac{1/d_i^{\beta}}{\sum_{i=1}^{n}(1/d_i^{\beta})}$, where β is the distance power that determines the degree to which nearer locations are preferred over more distant locations. For example, if $\beta = 1$, $w_i = \frac{1/d_i}{\sum_{i=1}^{n}(1/d_i)}$.

We can apply the IDW method with the `gstat()` function of **gstat** and the following arguments:

- `formula`: vble ~ 1 to have an intercept only model,
- `nmax`: number of neighbors is set equal to the total number of locations,
- `idp`: inverse distance power is set to `idp = 1` to have weights with $\beta = 1$.

Then, we use the `predict()` function to obtain the predictions and **tmap** to show the results (Figure 13.3).

```
library(gstat)
res <- gstat(formula = vble ~ 1, locations = d,
             nmax = nrow(d), # use all the neighbors locations
             set = list(idp = 1)) # beta = 1

resp <- predict(res, coop)
resp$x <- st_coordinates(resp)[,1]
resp$y <- st_coordinates(resp)[,2]
resp$pred <- resp$var1.pred

pred <- terra::rasterize(resp, grid, field = "pred", fun = "mean")
tm_shape(pred) + tm_raster(alpha = 0.6, palette = "viridis")
```

13.1.5   Nearest neighbors

In the nearest neighbors interpolation method, values at unsampled locations are estimated as the average of the values of the k closest sampled locations. Specifically,

$$\hat{Z}(\boldsymbol{s}_0) = \frac{\sum_{i=1}^{k} Z(\boldsymbol{s}_i)}{k},$$

where $\hat{Z}(\boldsymbol{s}_0)$ is the predicted value at $\boldsymbol{s}_0$, $Z(\boldsymbol{s}_i)$ is the observed value corresponding to neighbor $\boldsymbol{s}_i$, and k is the number of neighbors considered.

We can compute predictions using nearest neighbors interpolation with the gstat() function of **gstat**. Here, we consider the number of closest sampled locations equal to 5 by setting nmax = 5. Unlike the IDW method, in the nearest neighbors approach locations further away from the location where we wish to predict are assigned the same weights. Therefore, the inverse distance power idp is set equal to zero so all the neighbors are equally weighted.

Then, we use the predict() function to get predictions at unsampled locations given in coop. Figure 13.3 shows the map with the predictions.

```
library(gstat)
res <- gstat(formula = vble ~ 1, locations = d, nmax = 5,
             set = list(idp = 0))

resp <- predict(res, coop)
resp$x <- st_coordinates(resp)[,1]
resp$y <- st_coordinates(resp)[,2]
resp$pred <- resp$var1.pred

pred <- terra::rasterize(resp, grid, field = "pred", fun = "mean")
tm_shape(pred) + tm_raster(alpha = 0.6, palette = "viridis")
```

13.1.6 Ensemble approach

Predictions can also be obtained using an ensemble approach that combines the predictions obtained with several spatial interpolation methods. Specifically, if M is the number of interpolation methods considered, the predicted value $\hat{Z}(s_0)$ can be obtained as

$$\hat{Z}(s_0) = \sum_{i=1}^{M} \hat{Z}^{(i)}(s_0) \times w_i,$$

where $\hat{Z}^{(i)}(s_0)$ and w_i are, respectively, the prediction value and the weight corresponding to method i, with $i = 1, \ldots, M$. The weights can be chosen in different ways. For example, they can be proportional to some goodness-of-fit measure of each method and sum to 1.

Here, we use an ensemble approach to predict the price per square meter of apartments in Athens by combining the predictions of the three previous approaches (closest observation, IDW, nearest neighbors) using equal weights.

The predictions with the closest observation method are obtained using the Voronoi diagram as follows:

```
# Closest observation (Voronoi)
v <- terra::voronoi(x = terra::vect(d), bnd = map)
v <- st_as_sf(v)
p1 <- st_intersection(v, coop)$vble
```

The IDW approach can be applied with the gstat() function of **gstat** specifying **nmax** as the total number of locations and idp = 1.

```
# IDW
gs <- gstat(formula = vble ~ 1, locations = d, nmax = nrow(d),
            set = list(idp = 1))
p2 <- predict(gs, coop)$var1.pred
```

The nearest neighbors method is applied with the gstat() function specifying **nmax** as the number of neighbors and with equal weights (idp = 0).

```
# Nearest neighbors
nn <- gstat(formula = vble ~ 1, locations = d, nmax = 5,
            set = list(idp = 0))
p3 <- predict(nn, coop)$var1.pred
```

Finally, the ensemble predictions are obtained by combining the predictions of these three methods with equal weights.

```
# Ensemble (equal weights)
weights <- c(1/3, 1/3, 1/3)
p4 <- p1 * weights[1] + p2 * weights[2] + p3 * weights[3]
```

We create a map with the predictions by creating a raster with the predictions and using the **tmap** package (Figure 13.3).

```
resp <- data.frame(
x = st_coordinates(coop)[, 1],
y = st_coordinates(coop)[, 2],
pred = p4)

resp <- st_as_sf(resp, coords = c("x", "y"), crs = st_crs(map))

pred <- terra::rasterize(resp, grid, field = "pred", fun = "mean")
tm_shape(pred) + tm_raster(alpha = 0.6, palette = "viridis")
```

13.1.7 Cross-validation

We can assess the performance of each of the methods presented above using K-fold cross-validation and the root mean squared error (RMSE). First, we split the data in K parts. For each part, we use the remaining $K - 1$ parts (training data) to fit the model and that part (testing data) to predict. We compute the RMSE by comparing the testing and predicted data in each of the K parts:

$$RMSE = \left(\frac{1}{n_{test}} \sum_{i=1}^{n_{test}} (y_i^{test} - \widehat{y_i^{test}})^2 \right)^{1/2},$$

where y_i^{test} and $\widehat{y_i^{test}}$ are the observed and predicted values of observation i in the test set, and n_{test} is the number of observations in the test set. Finally, we average the $RMSE$ values obtained in each of the K parts.

Note that if K is equal to the number of observations n, this procedure is called leave-one-out cross-validation (LOOCV). That means that n separate data sets are trained on all of the data except one observation, and then prediction is made for that one observation.

Here, we assess the performance of each of the methods previously employed to predict the prices of apartments in Athens. We create training and testing sets by using the `dismo:kfold()` function of the **dismo** package (Hijmans et al., 2022) to randomly assign the observations to $K = 5$ groups of roughly equal size. For each group, we fit the model using the training data, and obtain predictions of the testing data. We calculate the RMSEs of each part and average the RMSEs to obtain a K-fold cross-validation estimate.

```
set.seed(123)

# Function to calculate the RMSE
RMSE <- function(observed, predicted) {
sqrt(mean((observed - predicted)^2))
}

# Split data in 5 sets
kf <- dismo::kfold(nrow(d), k = 5) # K-fold partitioning

# Vectors to store the RMSE values obtained with each method
rmse1 <- rep(NA, 5) # Closest observation
rmse2 <- rep(NA, 5) # IDW
rmse3 <- rep(NA, 5) # Nearest neighbors
rmse4 <- rep(NA, 5) # Ensemble
```

```
for(k in 1:5) {
# Split data in test and train
test <- d[kf == k, ]
train <- d[kf != k, ]

# Closest observation
v <- terra::voronoi(x = terra::vect(train), bnd = map)
v <- st_as_sf(v)
p1 <- st_intersection(v, test)$vble
rmse1[k] <- RMSE(test$vble, p1)

# IDW
gs <- gstat(formula = vble ~ 1, locations = train,
            nmax = nrow(train), set = list(idp = 1))
p2 <- predict(gs, test)$var1.pred
rmse2[k] <- RMSE(test$vble, p2)

# Nearest neighbors
nn <- gstat(formula = vble ~ 1, locations = train,
            nmax = 5, set = list(idp = 0))
p3 <- predict(nn, test)$var1.pred
rmse3[k] <- RMSE(test$vble, p3)

# Ensemble (weights are inverse RMSE so lower RMSE higher weight)
w <- 1/c(rmse1[k], rmse2[k], rmse3[k])
weights <- w/sum(w)
p4 <- p1 * weights[1] + p2 * weights[2] + p3 * weights[3]
rmse4[k] <- RMSE(test$vble, p4)
}
```

The RMSE values obtained in each of the 5 splits are shown below. We see the minimum average RMSE corresponds to the ensemble method.

```
# RMSE obtained for each of the 5 splits
data.frame(closest.obs = rmse1, IDW = rmse2,
           nearest.neigh = rmse3, ensemble = rmse4)
```

	closest.obs	IDW	nearest.neigh	ensemble
1	960.0	855.1	819.8	732.7
2	836.8	762.7	700.7	678.1
3	1038.5	962.8	867.1	868.9
4	1003.1	921.6	870.3	843.2
5	803.5	844.2	783.0	717.8

```
# Average RMSE over the 5 splits
data.frame(closest.obs = mean(rmse1), IDW = mean(rmse2),
           nearest.neigh = mean(rmse3), ensemble = mean(rmse4))


  closest.obs   IDW nearest.neigh ensemble
1       928.4 869.3         808.2    768.2
```

14

Kriging

Kriging (Matheron, 1963) is a spatial interpolation method used to obtain predictions at unsampled locations based on observed geostatistical data. This method originated in the field of mining geology and is named after South African mining engineer Danie G. Krige. Suppose that we have observed data $Z(s_1), \ldots, Z(s_n)$, and wish to predict the value of Z at an arbitrary location $s_0 \in D$. The Ordinary Kriging estimator of $Z(s_0)$ is defined as the linear unbiased estimator

$$\hat{Z}(s_0) = \sum_{i=1}^{n} \lambda_i Z(s_i)$$

that minimizes the mean squared prediction error defined as

$$E[(\hat{Z}(s_0) - Z(s_0))^2].$$

The Kriging weights are derived from the estimated spatial structure of the sampled data. Specifically, weights are obtained by first fitting a variogram model to the observed data, which helps us understand how the correlation between observation values changes with the distance between locations. Once the Kriging weights are obtained, they are applied to the known data values at the sampled locations to calculate the predicted values at unsampled locations. The Kriging weights reflect the spatial correlation of the data, which accounts for the geographical proximity and similarity of data points. Thus, observed locations that are correlated and near to the prediction locations are given more weight than those uncorrelated and farther apart. Weights also take into account the spatial arrangement of all observations, so clusterings of observations in oversampled areas are weighted less heavily since they contain less information than single locations.

Under several assumptions, Kriging predictions are best linear unbiased estimators. There are several types of Kriging differing by underlying assumptions and analytic goals. For example, Simple Kriging assumes the mean of the random field, $\mu(s)$, is known; Ordinary Kriging assumes a constant unknown mean, $\mu(s) = \mu$; and Universal Kriging can be used for data with an unknown non-stationary mean structure.

14.1 Kriging predictions of zinc concentrations

The **gstat** package (Pebesma and Graeler, 2022) has functionality to model, predict, and simulate geostatistical data. Here, we provide an example on how to predict zinc concentrations using samples collected near the river Meuse, The Netherlands. First, we load the data and create maps with the sample and the prediction locations. Then, we show how to fit a model variogram to the empirical variogram, and we use this fit to obtain predictions using Kriging. Finally, we create maps of the predictions and their associated uncertainty.

14.1.1 Data

The `meuse` data from the **sp** package contains zinc and other soil-heavy metal concentrations collected at locations in a flood plain of the river Meuse near Stein, The Netherlands (Figure 14.1). `meuse.grid` contains prediction grid locations for the `meuse` dataset (Figure 14.2). We convert the `meuse` and `meuse.grid` data frames to **sf** objects using the `st_as_sf()` function and setting the coordinate reference system to EPSG 28992. We create maps using the **mapview** package.

```
library(sp)
library(gstat)
library(sf)
library(mapview)

data(meuse)
data(meuse.grid)

meuse <- st_as_sf(meuse, coords = c("x", "y"), crs = 28992)
mapview(meuse, zcol = "zinc",  map.types = "CartoDB.Voyager")

meuse.grid <- st_as_sf(meuse.grid, coords = c("x", "y"),
                       crs = 28992)
mapview(meuse.grid,  map.types = "CartoDB.Voyager")
```

14.1.2 Variogram cloud

The variogram cloud shows half of all possible squared differences of data observation pairs against their separation distance h:

$$\frac{1}{2}(Z(s) - Z(s+h))^2$$

FIGURE 14.1: Zinc concentrations at locations in a flood plain of the river Meuse.

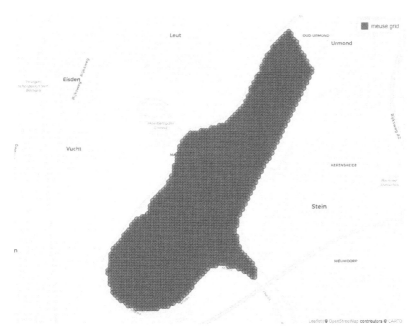

FIGURE 14.2: Prediction locations.

The `variogram()` function of **gstat** can be used to calculate the variogram cloud. In `variogram()`, we set argument `object` to `z ~ 1` if we wish to obtain the variogram for data `z`, or to a formula of a linear model with covariates if we wish the variogram for the residuals. Figure 14.3 shows the variogram cloud for the zinc data. This plot can be inspected to assess whether sample pairs closer to each other are more similar than pairs further apart.

```
vc <- variogram(log(zinc) ~ 1, meuse, cloud = TRUE)
plot(vc)
```

14.1.3   Sample variogram

Assuming isotropy, the sample variogram averages the variogram cloud values over distance intervals:

$$2\hat{\gamma}(h) = \frac{1}{|N(h)|} \sum_{N(h)} (Z(s_i) - Z(s_j))^2,$$

where $|N(h)|$ denotes the number of distinct pairs in

$$N(h) = \{(s_i, s_j) : ||s_i - s_j|| = h, \ i, j = 1, \ldots, n\}.$$

The `variogram()` function of **gstat** calculates the sample variogram from data or for residuals if a linear model has been specified. We can also specify the argument `width` indicating the width of distance intervals into which data point pairs are grouped to compute the estimates. Figure 14.3 shows the sample variogram for the zinc data.

```
v <- variogram(log(zinc) ~ 1, data = meuse)
plot(v)
```

14.1.4   Fitted variogram

The sample variogram obtained is a function of distance h estimated at discrete lags (e.g., $h(1), h(2), \ldots, h(k)$). Then, we can fit a variogram model to these estimated values,

$$\{(h(j), \ 2\hat{\gamma}(h(j))) : j = 1, 2, \ldots, k\}.$$

The `vgm()` function of **gstat** generates a variogram model. This function has arguments `model` with the model type (e.g., `Sph` for spherical and `Exp` for

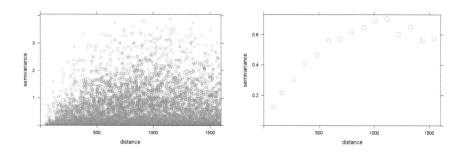

FIGURE 14.3: Variogram cloud (left) and sample variogram (right).

exponential), `nugget`, `psill` (partial sill, which is the sill minus the nugget) and `range`. Details on the construction of the variogram models are given in Section 4.3 of the **gstat** user's manual[1]. A list of available models can be printed with `vgm()` and visualized with `show.vgms()` (Figure 14.4).

`vgm()`

	short	long
1	Nug	Nug (nugget)
2	Exp	Exp (exponential)
3	Sph	Sph (spherical)
4	Gau	Gau (gaussian)
5	Exc	Exclass (Exponential class/stable)
6	Mat	Mat (Matern)
7	Ste	Mat (Matern, M. Stein's parameterization)
8	Cir	Cir (circular)
9	Lin	Lin (linear)
10	Bes	Bes (bessel)
11	Pen	Pen (pentaspherical)
12	Per	Per (periodic)
13	Wav	Wav (wave)
14	Hol	Hol (hole)
15	Log	Log (logarithmic)
16	Pow	Pow (power)
17	Spl	Spl (spline)
18	Leg	Leg (Legendre)
19	Err	Err (Measurement error)
20	Int	Int (Intercept)

---

[1]https://www.gstat.org/gstat.pdf

```
show.vgms(par.strip.text = list(cex = 0.75))
```

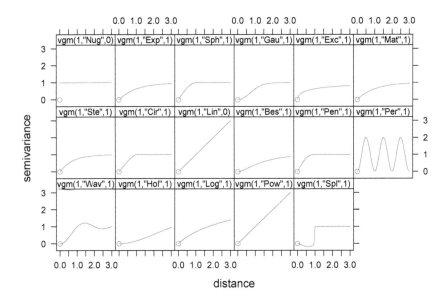

FIGURE 14.4: Available variogram models in **gstat**.

The `fit.variogram()` function fits a variogram model to a sample variogram. Arguments of this function include `object` with the sample variogram, and `model` with the variogram model which is an output of `vgm()` with arguments `nugget`, `psill`, and `range` denoting initial values of the iterative fitting algorithm. The fitting method is specified in `fit.method` and can be ordinary (unweighted) or weighted least squares using a weighting scheme that gives most weight to early lags and downweights those lags with a small number of pairs. Thus, if $\{2\gamma(h; \boldsymbol{\lambda})\}$ is a model depending on parameters $\boldsymbol{\lambda}$, and w_j are weights associated to lag $h(j)$, the method of weighted least squares chooses the value of $\boldsymbol{\lambda}$ that minimizes

$$\sum_{j=1}^{k} w_j \left[2\hat{\gamma}(h(j)) - 2\gamma(h(j); \boldsymbol{\lambda})\right]^2.$$

By inspecting the sample variogram of the zinc data, we may think that a good model for the variogram would be spherical with initial values for the partial sill equal to 0.5, range equal to 900, and nugget equal to 0.1. Figure 14.5 (left) shows the sample variogram v calculated with `variogram()`, and the variogram generated with **vgm()** using a spherical model, and with partial sill, range, and nugget equal to our initial guess values. This plot allows us to assess our initial model.

```
vinitial <- vgm(psill = 0.5, model = "Sph",
                range = 900, nugget = 0.1)
plot(v, vinitial, cutoff = 1000, cex = 1.5)
```

Then, we fit a variogram providing the sample variogram and choosing a spherical model (**model = "Sph"**), and the initial values for the partial sill, range, and the nugget for the iterative fitting algorithm. Figure 14.5 (right) shows the sample and the fitted variogram.

```
fv <- fit.variogram(object = v,
                    model = vgm(psill = 0.5, model = "Sph",
                                range = 900, nugget = 0.1))
fv
plot(v, fv, cex = 1.5)
```

```
  model    psill range
1   Nug 0.05066     0
2   Sph 0.59061   897
```

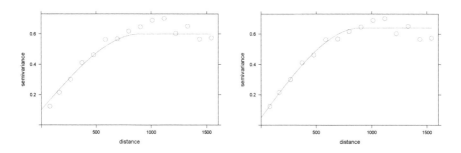

FIGURE 14.5: Sample variogram and variogram with initial parameters (left), and fitted model (right).

14.1.5   Kriging

Kriging uses the fitted variogram values **fv** to obtain predictions at unsampled locations. Here, we use the **gstat()** function of **gstat** to compute the Kriging predictions.

```
library(ggplot2)
library(viridis)

k <- gstat(formula = log(zinc) ~ 1, data = meuse, model = fv)
```

Then, we obtain predictions at the unsampled locations given by `meuse.grid` with the `predict.gstat()` function.

```
kpred <- predict(k, meuse.grid)
```

The returned object of `predict()` contains the predictions (`var1.pred`) and their variance (`var1.var`) which allow us to quantify the uncertainty of these predictions. Figure 14.6 shows maps with the predictions and variance values. Note that the Kriging predictions are calculated using weights that depend on the variogram estimates, and they are therefore sensitive to misspecification of the variogram model.

```
ggplot() + geom_sf(data = kpred, aes(color = var1.pred)) +
  geom_sf(data = meuse) +
  scale_color_viridis(name = "log(zinc)") + theme_bw()

ggplot() + geom_sf(data = kpred, aes(color = var1.var)) +
  geom_sf(data = meuse) +
  scale_color_viridis(name = "variance") + theme_bw()
```

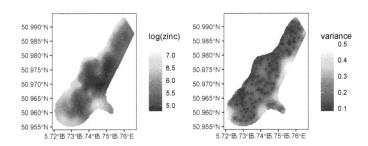

FIGURE 14.6: Predictions (left) and variance (right) obtained with Kriging.

15

Model-based geostatistics

Model-based geostatistics can be used to analyze spatial data related to an underlying spatially continuous phenomenon that have been collected at a finite set of locations. Model-based geostatistics employs statistical models to capture the spatial correlation structure in the data, enabling rigorous statistical inference, and facilitating the production of spatial predictions along with uncertainty measures of the phenomenon of interest (Diggle et al., 1998).

Assuming Gaussian data observed at a set of n locations, $\{Y_1, \ldots, Y_n\}$, we can consider the following model to obtain predictions at unsampled locations:

$$Y_i | S(\boldsymbol{s}_i) \sim N(\mu + S(\boldsymbol{s}_i), \tau^2), i = 1, \ldots, n.$$

Here, μ is a constant mean effect, and $S(\cdot)$ is a zero-mean spatial Gaussian field. This model can be extended to situations in which the stochastic variation in the data is not Gaussian, as well as to include covariates and other random effects to account for other types of variability.

Inference in model-based geostatistics can be performed using the INLA and the stochastic partial differential equation (SPDE) approaches which provide a computationally efficient alternative to MCMC methods (Lindgren and Rue, 2015). Briefly, this involves solving a SPDE on a discrete mesh of points and interpolating to obtain a continuous solution across the spatial domain (Krainski et al., 2019) which is calculated using INLA (Rue et al., 2009).

Model-based geostatistics using INLA and SPDE has been employed for spatial prediction in a wide range of applications including air pollution in Italy (Cameletti et al., 2013), leptospirosis in Brazil (Hagan et al., 2016), and malaria in Mozambique (Moraga et al., 2021). Model-based geostatistics also provides a flexible approach to combine multiple data available at different spatial resolutions to get better predictions than the ones obtained using just one type of data (Zhong and Moraga, 2023). For example, Moraga et al. (2017) demonstrate how to integrate air pollution measures obtained at a collection of monitoring stations, and aggregated at cells of a regular grid derived from satellites to obtain predictions at a continuous surface and improve decision-making. Moreover, model-based geostatistics can also account for preferential sampling that may occur when the spatial phenomenon of interest and the sampling locations exhibit stochastic dependence (Diggle et al.,

2010). Specifically, spatially misaligned data can be combined by assuming a common spatial random field underlying all observations. In the integration of data, preferential sampling can be taken into account by assuming a shared spatial random process by both the measured observations and the intensity of the point process that originates the locations, with a parameter controlling the degree of preferential sampling (Zhong et al., 2023; Ribeiro Amaral et al., 2023b).

In this chapter, we introduce the SPDE approach, and show how to specify, fit, and interpret a geostatistical model to predict air pollution in the USA using INLA and SPDE. Additional examples on how to implement the INLA and SPDE approaches to fit geostatistical models are provided in Krainski et al. (2019) and Moraga (2019). Specifically, Moraga (2019) demonstrates how to fit both spatial and spatio-temporal models to predict malaria prevalence in The Gambia, precipitation in Brazil, and air pollution in Spain.

15.1   The SPDE approach

The stochastic partial differential equation (SPDE) approach implemented in the **R-INLA** package provides a flexible and computationally efficient way to model geostatistical data and make predictions at unsampled locations (Lindgren and Rue, 2015). We assume that underlying the observed data, there is a spatially continuous variable that can be modeled using a Gaussian random field (GRF) with Matérn covariance function which is defined as

$$\text{Cov}(x(\boldsymbol{s}_i), x(\boldsymbol{s}_j)) = \frac{\sigma^2}{2^{\nu-1}\Gamma(\nu)}(\kappa||\boldsymbol{s}_i - \boldsymbol{s}_j||)^\nu K_\nu(\kappa||\boldsymbol{s}_i - \boldsymbol{s}_j||).$$

Here, σ^2 denotes the marginal variance of the spatial field. $K_\nu(\cdot)$ refers to the modified Bessel function of second kind and order $\nu > 0$. The integer value of ν determines the smoothness of the field and is typically fixed since it is difficult to estimate in applications. $\kappa > 0$ is related to the range ρ, which represents the distance at which the correlation between two points becomes approximately 0. Specifically, $\rho = \sqrt{8\nu}/\kappa$, and at this distance the spatial correlation is close to 0.1 (Cameletti et al., 2013).

As shown in Whittle (1963), a GRF with a Matérn covariance matrix can be represented as a solution of the following continuous domain SPDE:

$$(\kappa^2 - \Delta)^{\alpha/2}(\tau x(\boldsymbol{s})) = \mathcal{W}(\boldsymbol{s}).$$

Here, $x(s)$ represents a GRF, and $\mathcal{W}(s)$ is a Gaussian spatial white noise process. The parameter α controls the smoothness exhibited by the GRF, τ controls its variance, and $\kappa > 0$ is a scale parameter. The Laplacian Δ is defined as $\sum_{i=1}^{d} \frac{\partial^2}{\partial x_i^2}$, where d is the dimension of the spatial domain.

The parameters of the Matérn covariance function and the SPDE are related as follows. The smoothness parameter ν of the Matérn covariance function is expressed as $\nu = \alpha - \frac{d}{2}$, and the marginal variance σ^2 is related to the SPDE through

$$\sigma^2 = \frac{\Gamma(\nu)}{\Gamma(\alpha)(4\pi)^{d/2}\kappa^{2\nu}\tau^2}.$$

In the case where $d = 2$ and $\nu = 1/2$, which corresponds to the exponential covariance function, the parameter $\alpha = \nu + d/2 = 1/2 + 1 = 3/2$. In the **R-INLA** package, the default value is $\alpha = 2$, although options within the range $0 \leq \alpha < 2$ are also available.

The Finite Element method can be used to find an approximate solution to the SPDE. This method involves dividing the spatial domain into a set of non-intersecting triangles, creating a triangulated mesh with n nodes and n basis functions. The basis functions, denoted as $\psi_k(\cdot)$, are piecewise linear functions on each triangle. They take the value of 1 at vertex k, and 0 at all other vertices.

Then, the continuously indexed Gaussian field x is represented as a discretely indexed Gaussian Markov random field (GMRF) by a sum of basis functions defined on the triangulated mesh

$$x(s) = \sum_{k=1}^{n} \psi_k(s)x_k,$$

where n is the number of vertices of the triangulation, $\psi_k(\cdot)$ represents the piecewise linear basis functions, and $\{x_k\}$ denote zero-mean Gaussian distributed weights.

The joint distribution of the weight vector is assigned a Gaussian distribution represented as $x = (x_1, \ldots, x_n) \sim N(\mathbf{0}, \mathbf{Q}^{-1}(\tau, \kappa))$. This distribution approximates the solution $x(s)$ of the SPDE at the mesh nodes. The basis functions transform the approximation $x(s)$ from the mesh nodes to the other spatial locations of interest.

Projection matrix

The SPDE approach can be implemented with **R-INLA** by creating a projection matrix A that projects the GRF from the observations to the vertices of the triangulated mesh. The projection matrix A has a number of rows equal to the number of observations, and a number of columns equal to the number of vertices of the mesh. Each row i of A corresponds to an observation at location s_i, and may have up to three non-zero values in the columns that correspond to the vertices of the triangle containing the location. If s_i lies within the triangle, these values are equal to the barycentric coordinates. In other words, they are proportional to the areas of each of the three subtriangles formed by the location s_i and the triangle's vertices, and they sum to 1. If s_i coincides with a vertex of the triangle, row i has just one non-zero value equal to 1 in the column that corresponds to that vertex.

For example, the projection matrix below has n rows as the number of observations, and G columns as the number of vertices of the triangulated mesh. The first row of the matrix corresponds to an observation located at the vertex in the third position. The second and last rows correspond to observations located within triangles.

$$
A = \begin{bmatrix} A_{11} & A_{12} & A_{13} & \cdots & A_{1G} \\ A_{21} & A_{22} & A_{23} & \cdots & A_{2G} \\ \vdots & \vdots & \vdots & \ddots & \vdots \\ A_{n1} & A_{n2} & A_{n3} & \cdots & A_{nG} \end{bmatrix} = \begin{bmatrix} 0 & 0 & 1 & \cdots & 0 \\ A_{21} & A_{22} & 0 & \cdots & A_{2G} \\ \vdots & \vdots & \vdots & \ddots & \vdots \\ A_{n1} & A_{n2} & A_{n3} & \cdots & 0 \end{bmatrix}
$$

Figure 15.1 shows a location s within one of the triangles of a triangulated mesh. The value of the process $Z(\cdot)$ at s is determined through a weighted average of the process values at the triangle's vertices (Z_1, Z_2, Z_3). These weights are calculated by dividing the areas of the subtriangles (T_1, T_2, T_3) by the area of the larger triangle that contains s (T):

$$
Z(s) \approx \frac{T_1}{T} Z_1 + \frac{T_2}{T} Z_2 + \frac{T_3}{T} Z_3.
$$

Thus, the value $Z(s)$ at a location within a mesh triangle can be obtained by projecting the plane formed by the triangle vertices weights at location s.

15.2   Air pollution prediction

In this section, we show how to fit a geostatistical model to predict fine particulate matter PM2.5 in the USA using the INLA and SPDE approaches.

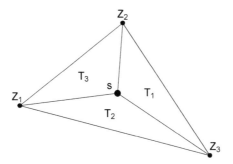

FIGURE 15.1: Triangle of a triangulated mesh.

15.2.1 Observed PM2.5 values in the USA

Annual averages of PM2.5 concentration levels recorded at 1429 monitoring stations from the United States Environmental Protection Agency[1] in 2022 are in the PM25USA2022.csv file that can be downloaded from this website[2]. We use the read.csv() function to read the data which contains the longitude and latitude values of the monitoring stations, and the recorded PM2.5 values in micrograms per cubic meter. Then, we use the st_as_sf() function to transform the data.frame obtained to a sf object with geographic CRS given by EPSG code 4326.

```
library(sf)
f <- file.path("https://www.paulamoraga.com/book-spatial/",
               "data/PM25USA2022.csv")
d <- read.csv(f)
d <- st_as_sf(d, coords = c("longitude", "latitude"))
st_crs(d) <- "EPSG:4326"
```

We then obtain the map of the USA with the ne_countries() function of **rnaturalearth**. We use st_crop() to remove Alaska and other areas that are outside the region comprised by longitude values (–130, 60) and latitude values (18, 72).

```
library(rnaturalearth)
map <- ne_countries(type = "countries",
                    country = "United States of America",
                    scale = "medium", returnclass = "sf")
map <- st_crop(map, xmin = -130, xmax = -60, ymin = 18, ymax = 72)
```

[1]https://www.epa.gov/
[2]https://www.paulamoraga.com/book-spatial/data/PM25USA2022.csv

We then keep the 1366 monitoring stations locations that are within the map
by using the `st_filter()` function.

```
d <- st_filter(d, map)
nrow(d)
```

```
[1] 1366
```

Figure 15.2 shows a map with the PM2.5 observed values.

```
library(ggplot2)
library(viridis)
ggplot() + geom_sf(data = map) +
  geom_sf(data = d, aes(col = value)) +
  scale_color_viridis()
```

15.2.2   Prediction data

Here, we construct a matrix `coop` with the locations where the air pollution
levels will be predicted. First, we create a raster grid with 100×100 cells
covering the map using the `rast()` function of **terra**. Then, we obtain the
coordinates of the cells with the `xyfromCell()` function of **terra**.

```
library(sf)
library(terra)

# raster grid covering map
grid <- terra::rast(map, nrows = 100, ncols = 100)
# coordinates of all cells
xy <- terra::xyFromCell(grid, 1:ncell(grid))
```

Then, we use the `st_as_sf()` function to create a **sf** object with the coordinates of the prediction locations by specifying the coordinates as a data frame,
the name of the coordinates, and the CRS. We obtain the indices of the point
coordinates that are within the map with `st_intersects()` setting `sparse =`
`FALSE`. We will later use these indices to identify the prediction locations. We
also obtain the point coordinates that are within the map with `sf_filter()`.
Figure 15.2 shows the prediction locations.

```
# transform points to a sf object
dp <- st_as_sf(as.data.frame(xy), coords = c("x", "y"),
              crs = st_crs(map))
```

```
# indices points within the map
indicespointswithin <- which(st_intersects(dp, map,
                                            sparse = FALSE))

# points within the map
dp <- st_filter(dp, map)

# plot
ggplot() + geom_sf(data = map) +
  geom_sf(data = dp)
```

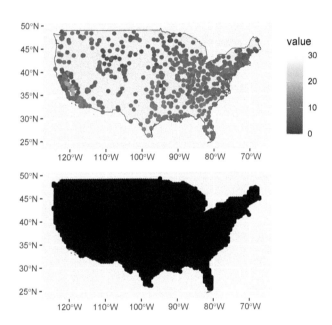

FIGURE 15.2: Map with the PM2.5 observed values (top) and prediction locations (bottom).

15.2.3 Covariates

In our model, we use average temperature and precipitation as covariates. Monthly values of these variables globally can be obtained with the worldclim_global() function of **geodata**.

```
library(geodata)
covtemp <- worldclim_global(var = "tavg", res = 10,
```

```
                         path = tempdir())
covprec <- worldclim_global(var = "prec", res = 10,
                         path = tempdir())
```

After downloading the data, we compute the averages over months and extract the values at the observation and prediction locations with the extract() function of **terra**.

```
# Extract at observed locations
d$covtemp <- extract(mean(covtemp), st_coordinates(d))[, 1]
d$covprec <- extract(mean(covprec), st_coordinates(d))[, 1]
# Extract at prediction locations
dp$covtemp <- extract(mean(covtemp), st_coordinates(dp))[, 1]
dp$covprec <- extract(mean(covprec), st_coordinates(dp))[, 1]
```

Figure 15.3 shows maps of the temperature and precipitation covariates at the observation locations created with the **ggplot2** and **patchwork** packages.

```
library("patchwork")
p1 <- ggplot() + geom_sf(data = map) +
  geom_sf(data = d, aes(col = covtemp)) +
  scale_color_viridis()
p2 <- ggplot() + geom_sf(data = map) +
  geom_sf(data = d, aes(col = covprec)) +
  scale_color_viridis()
p1 / p2
```

15.2.4   Transforming coordinates to UTM

The data we are dealing with have a geographic CRS that references locations using longitude and latitude values. In order to work with kilometers instead of degrees, we use st_transform() to transform the CRS of the sf objects with the data corresponding to the observed (d) and the prediction (dp) locations from geographic to a projected CRS. Specifically, we use the Mercator projection that is given by EPSG code 3857 and use kilometers as units. To do that, we use the projection given by st_crs("EPSG:3857")$proj4string replacing +units=m by +units=km.

```
st_crs("EPSG:3857")$proj4string
projMercator<-"+proj=merc +a=6378137 +b=6378137 +lat_ts=0 +lon_0=0
+x_0=0 +y_0=0 +k=1 +units=km +nadgrids=@null +wktext +no_defs"
```

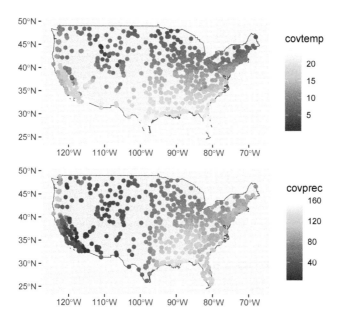

FIGURE 15.3: Temperature and precipitation measurements at the observation locations.

```
d <- st_transform(d, crs = projMercator)
dp <- st_transform(dp, crs = projMercator)
```

15.2.5   Coordinates of observed and prediction locations

After transforming the data, we use the `st_coordinates()` function to create matrices with the projected coordinates of the observed and prediction locations.

```
# Observed coordinates
coo <- st_coordinates(d)

# Predicted coordinates
coop <- st_coordinates(dp)
```

15.2.6   Model

Now we specify the model that we use to predict PM2.5 values at unsampled locations. We assume that Y_i, the PM2.5 values measured at locations $i = 1, \ldots, n$, can be modeled as

$$Y_i \sim N(\mu_i, \sigma^2),$$

$$\mu_i = \beta_0 + \beta_1 \times \text{temp}_i + \beta_2 \times \text{prec}_i + S(\boldsymbol{x}_i),$$

where β_0 is the intercept, and β_1 and β_2 are, respectively, the coefficients of temperature and precipitation. $S(\cdot)$ is a spatial random effect that is modeled as a zero-mean Gaussian process with Matérn covariance function.

15.2.7   Mesh construction

To fit the model using the SPDE approach, we first create a triangulated mesh covering the study region where we approximate the Gaussian random field as a Gaussian Markov random field. INLA produces good approximations by using a fine mesh consisting of very small triangles and with a large separation distance between the locations and the mesh boundary to avoid boundary effects by which the variance is increased near the boundary. In some applications the use of such a fine mesh could be computationally intensive, and we usually work with meshes that still produce good approximations consisting of an inner region with small triangles where precision is needed, and an outer extension with larger triangles where accurate approximations are not needed.

Here, we create the mesh with the `inla.mesh.2d()` function of **R-INLA**. We pass as arguments `loc = coo` with the location coordinates, and `max.edge = c(200, 500)` with the maximum allowed triangle edge lengths in the region and the extension to have smaller triangles within the region than in the extension. We also specify `cutoff = 1` with the minimum allowed distance between points to avoid building many small triangles in areas where locations are located close to each other (Figure 15.4). The number of vertices of the mesh can be obtained with `mesh$n`, and the mesh can be plotted as follows.

```
library(INLA)
summary(dist(coo)) # summary of distances between locations
```

```
   Min. 1st Qu.  Median   Mean 3rd Qu.   Max.
      0    1107    1966   2242    3311   6318
```

```
mesh <- inla.mesh.2d(loc = coo, max.edge = c(200, 500),
                     cutoff = 1)
mesh$n
```

```
[1] 3711
```

```
plot(mesh)
points(coo, col = "red")
axis(1)
axis(2)
```

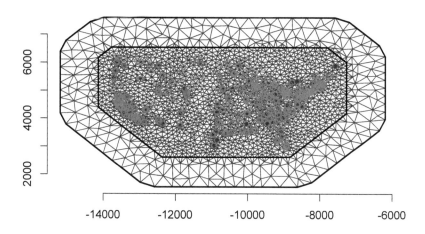

Constrained refined Delaunay triangulation

FIGURE 15.4: Mesh used in the SPDE approach. Locations of PM2.5 measurements are depicted as red points.

15.2.8 Building the SPDE on the mesh

Then, we use the `inla.spde2.matern()` function to build the SPDE model. This function has parameters `mesh` with the triangulated mesh constructed and `constr = TRUE` to impose an integrate-to-zero constraint. Moreover, we set the smoothness parameter ν equal to 1. In the spatial case $d = 2$ and $\alpha = \nu + d/2 = 2$.

```
spde <- inla.spde2.matern(mesh = mesh, alpha = 2, constr = TRUE)
```

15.2.9   Index set

Then, we create an index set for the SPDE model using the
`inla.spde.make.index()` function, where we provide the effect name
(s) and the number of vertices in the SPDE model (`spde$n.spde`). This
function generates a list with the vector s ranging from 1 to `spde$n.spde`.
Additionally, it creates two vectors, `s.group` and `s.repl`, containing all
elements set to 1 and lengths equal to the number of mesh vertices.

```
indexs <- inla.spde.make.index("s", spde$n.spde)
lengths(indexs)
```

```
      s s.group  s.repl
   3711    3711    3711
```

15.2.10   Projection matrix

We use the `inla.spde.make.A()` function of **R-INLA** passing the mesh
(`mesh`) and the coordinates (`coo`) to easily construct a projection matrix A that
projects the spatially continuous Gaussian random field from the observations
to the mesh nodes.

```
A <- inla.spde.make.A(mesh = mesh, loc = coo)
```

We can see the projection matrix A has a number of rows equal to the number
of observations, and a number of columns equal to the number of vertices of
the mesh. We also see the elements of each row of A sum to 1.

```
# dimension of the projection matrix
dim(A)
```

```
[1] 1366 3711
```

```
# number of observations
nrow(coo)
```

```
[1] 1366
```

```
# number of vertices of the triangulation
mesh$n
```

```
[1] 3711
```

```
# elements of each row sum to 1
# rowSums(A)
```

We also create a projection matrix for the prediction locations.

```
Ap <- inla.spde.make.A(mesh = mesh, loc = coop)
```

15.2.11   Stack with data for estimation and prediction

We now create a stack with the data for estimation and prediction that organizes data, effects, and projection matrices. We create stacks for estimation (`stk.e`) and prediction (`stk.p`) using `tag` to identify the type of data, `data` with the list of data vectors, `A` with the projection matrices, and `effects` with a list of fixed and random effects. First, we create a stack named `stk.e` containing the data for estimation which is tagged with the string `"est"`. In `data`, we specify the response vector with the observed PM2.5 values. The projection matrix is giving in argument `A`, which is a list where the second element is the projection matrix for the random effects (`A`) and the first element is set to 1 to indicate that the fixed effects are directly mapped one-to-one to the response. To define the effects, we pass a list containing the fixed and random effects. The fixed effects are a `data.frame` consisting of an intercept (`b0`) and covariates temperature (`covtemp`) and precipitation (`covprec`). The random effect is represented by the spatial Gaussian random field `s` containing a list with the indices of the SPDE object (`indexs`). Additionally, we construct another stack called `stk.p` for prediction, which is labeled with the tag `"pred"`. The data, projection matrix and effects are specified for the prediction locations. The response vector in the argument `data` of this stack is set to a list with `NA` because these are values we want to predict. Finally, we combine `stk.e` and `stk.p` into a single full stack named `stk.full`.

```
# stack for estimation stk.e
stk.e <- inla.stack(tag = "est",
data = list(y = d$value), A = list(1, A),
effects = list(data.frame(b0 = rep(1, nrow(A)),
covtemp = d$covtemp, covprec = d$covprec),
s = indexs))
```

```
# stack for prediction stk.p
stk.p <- inla.stack(tag = "pred",
data = list(y = NA), A = list(1, Ap),
effects = list(data.frame(b0 = rep(1, nrow(Ap)),
covtemp = dp$covtemp, covprec = dp$covprec),
s = indexs))

# stk.full has stk.e and stk.p
stk.full <- inla.stack(stk.e, stk.p)
```

15.2.12   Model formula and inla() call

Then, we specify the formula by including the response variable, the ~ symbol, and the fixed and random effects. In the formula, we eliminate the intercept by adding 0 and include it as a covariate term by adding b0. This step ensures that all covariate terms are properly captured within the projection matrix.

```
formula <- y ~ 0 + b0 + covtemp + covprec + f(s, model = spde)
```

Finally, we call inla() by specifying the formula, family, stack with the data, and options. We set control.predictor = list(compute = TRUE) and control.compute = list(return.marginals.predictor = TRUE) to compute and return the marginals for the linear predictor.

```
res <- inla(formula, family = "gaussian",
        data = inla.stack.data(stk.full),
        control.predictor = list(compute = TRUE,
                                  A = inla.stack.A(stk.full)),
        control.compute = list(return.marginals.predictor = TRUE))
```

15.2.13   Results

A summary of the results can be inspected with summary(res). The object res$summary.fixed provides the mean and quantiles of the posterior distribution of the intercept and coefficients of the covariates.

```
res$summary.fixed
```

	mean	sd	0.025quant	0.5quant
b0	3.88493	0.281396	3.326782	3.886731
covtemp	0.23917	0.019948	0.200858	0.238858

```
covprec 0.00273 0.003066  -0.003416 0.002772
         0.975quant       mode        kld
b0          4.432755 3.886717 1.730e-08
covtemp     0.279225 0.238854 5.777e-08
covprec     0.008634 0.002772 4.847e-08
```

We observe the coefficient of temperature is $\hat{\beta}_1 = 0.239$ with a 95% credible interval equal to (0.201, 0.279). The coefficient of precipitation is $\hat{\beta}_2 = 0.003$ with a 95% credible interval equal to (−0.003, 0.009). Thus, temperature is significantly associated with PM2.5, whereas precipitation is not significant.

15.2.14   Mapping predicted PM2.5 values

The res$summary.fitted.values object contains the posterior mean and quantiles of the fitted values. We can obtain the indices corresponding to the prediction locations by using the inla.stack.index() function passing the full stack and tag = "pred". Then, we retrieve the column "mean" with the posterior mean, and columns "0.025quant" and "0.975quant" with lower and upper limits of 95% credible intervals denoting the uncertainty of the predictions.

```
index <- inla.stack.index(stack = stk.full, tag = "pred")$data
pred_mean <- res$summary.fitted.values[index, "mean"]
pred_ll <- res$summary.fitted.values[index, "0.025quant"]
pred_ul <- res$summary.fitted.values[index, "0.975quant"]
```

We assign the predicted values to their corresponding cells within the map that are in the object grid that contains the prediction locations.

```
grid$mean <- NA
grid$ll <- NA
grid$ul <- NA

grid$mean[indicespointswithin] <- pred_mean
grid$ll[indicespointswithin] <- pred_ll
grid$ul[indicespointswithin] <- pred_ul

summary(grid) # negative values for the lower limit
```

```
      mean              ll               ul
Min.   : 2      Min.   : 0       Min.   : 3
1st Qu.: 6      1st Qu.: 4       1st Qu.: 7
Median : 7      Median : 5       Median : 8
Mean   : 7      Mean   : 5       Mean   : 8
```

```
3rd Qu.: 8       3rd Qu.: 6       3rd Qu.:10
Max.    :15       Max.    :13       Max.    :17
NA's    :4189     NA's    :4189     NA's    :4189
```

Then, we plot the posterior mean and 95% credible intervals of the predicted PM2.5 values with the `levelplot()` function of **rasterVis** package. Figure 15.5 depicts maps with the spatial pattern of the predicted PM2.5 levels as well as their associated uncertainty.

```
library(rasterVis)
levelplot(grid, layout = c(1, 3),
names.attr = c("Mean", "2.5 percentile", "97.5 percentile"))
```

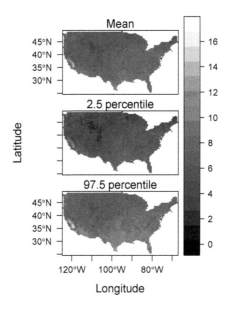

FIGURE 15.5: Posterior mean and lower and upper limits of uncertainty intervals of PM2.5.

15.2.15   Exceedance probabilities

We can also obtain probabilities that PM2.5 exceed a specific threshold level with the `inla.pmarginal()` function. Specifically, we calculate the probabilities that PM2.5 levels exceed 10 micrograms per cubic meter. That is, P(PM2.5 > 10) = 1 − P(PM2.5 ≤ 10).

```
excprob <- sapply(res$marginals.fitted.values[index],
FUN = function(marg){1-inla.pmarginal(q = 10, marginal = marg)})
```

Then, we add the exceedance probabilities as a layer in `grid`, and we plot it with `levelplot()`. In `levelplot()`, we set `margin = FALSE` to hide the marginal graphics of the column and row summaries of the raster object. Figure 15.6 shows the probabilities that PM2.5 levels exceed 10 micrograms per cubic meter. We observe high probabilities in the west coast and south part of the country.

```
grid$excprob <- NA
grid$excprob[indicespointswithin] <- excprob

levelplot(grid$excprob, margin = FALSE)
```

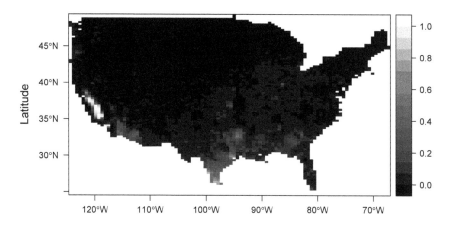

FIGURE 15.6: Probability that PM2.5 levels exceed 10 micrograms per cubic meter.

16

Methods assessment

The predictive performance of a spatial interpolation method can be assessed in several ways. For example, we can compare the observations and the predictions made at a set of locations using the Mean Absolute Error (MAE), the Root Mean Squared Error (RMSE), and the 95% Coverage Probability (CP).

Let y_i and $\hat{y}_i$ be the observed and predicted values, respectively, at locations x_i, $i = 1, \ldots, m$. We can calculate the Mean Absolute Error as

$$MAE = \frac{1}{m} \sum_{i=1}^{m} |y_i - \hat{y}_i|,$$

and the Root Mean Squared Error as

$$RMSE = \left(\frac{1}{m} \sum_{i=1}^{m} (y_i - \hat{y}_i)^2 \right)^{1/2}.$$

The 95% coverage probability is the proportion of times that the observed values are within their corresponding 95% prediction intervals and can be calculated as

$$\frac{1}{m} \sum_{i=1}^{m} I \left(y_i \in PI_i^{95\%} \right),$$

where $I \left(y_i \in PI_i^{95\%} \right)$ is an indicator function that takes the value 1 if y_i is inside its 95% prediction interval $PI_i^{95\%}$, and 0 otherwise.

If the spatial interpolation method provides prediction distributions, the Continuous Ranked Probability Score (CRPS) can be used to compare observations and predictions accounting for the uncertainty (Matheson and Winkler, 1976). The CRPS for observation y is a score function that compares the Cumulative Distribution Function (CDF) of the prediction distribution (F) with the degenerate CDF of the observation ($1\{u \geq y\}$):

$$CRPS(F, y) = \int_{-\infty}^{+\infty} (F(u) - 1\{u \geq y\})^2 \, du.$$

Figure 16.1 shows a plot representing the CRPS for a given observation. Note that a good method would yield a predicted distribution close to the observed value. Therefore, the preferred method would be one where the squared area between the two CDFs is small.

The CRPS for a set of observations can be calculated by aggregating the CRPS of the individual observations using an average or weighted average. A perfect CRPS score is equal to 0. Note that the CRPS reduces to the MAE if the predicted distribution is just a point estimate and not a distribution.

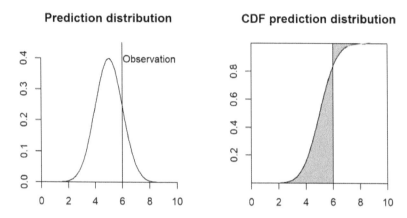

FIGURE 16.1: Left: Observation and prediction distribution. Right: Continuous Ranked Probability Score (CRPS) represented as the gray area between the cumulative distribution functions.

16.1 Cross-validation

The performance indices presented above can be computed using a new dataset or by splitting an existing dataset into a training dataset to fit the model and a testing dataset for validation. In cross-validation, the data is randomly split into several disjoint folds. Then, each fold is put aside in turn and used to evaluate the predictions obtained from a model fitted on the remaining folds. This procedure is called K-fold cross-validation if the data is split into K folds, and leave-one-out cross-validation (LOOCV) if each fold consists of only one observation.

In applications where spatial autocorrelation is present, randomly splitting the data into training and testing datasets may lead to an overestimation of the predictive performance since the characteristics of the testing and training datasets could be similar. In these situations, better performance measures may be used by employing spatial cross-validation and probability sampling.

Spatial cross-validation generates training and testing locations that are enough separated to provide independent datasets. For example, the **blockCV** package (Valavi et al., 2023) provides a range of functions to generate spatially and environmentally separated datasets for spatial K-fold and LOOCV. The package includes functions to construct spatial and buffer blocks, and to allocate them to cross-validation folds. It also has functionality to assess the level of spatial autocorrelation to help select the blocks and the buffer sizes appropriately.

Note also that spatial cross-validation techniques may generate training datasets where geographic and therefore covariate space is excluded causing under-representation of environmental conditions similar to those at the validation locations. Probability sampling can be used in these situations to obtain a more representative set of locations and avoid extrapolation problems (Wadoux et al., 2021).

Part IV

Spatial point patterns

17

Spatial point patterns

Spatial point patterns are countable sets of points that arise as realizations of stochastic spatial point processes taking values in a planar region $A \subset \mathbb{R}^2$. A spatial point pattern can be denoted as $\{\boldsymbol{x}_1, \boldsymbol{x}_2, \ldots, \boldsymbol{x}_{N(A)}\}$, where $N(A)$ is the number of points in A. Note that $N(A)$ is a random variable. Therefore, different realizations of the spatial point process may result in both different numbers and locations of points (Baddeley et al., 2016). We often refer to the points in the point pattern as events to distinguish them from arbitrary points in the plane.

Spatial point patterns arise in many domains. Examples include locations of individuals with a certain disease in a city (Moraga and Montes, 2011; Ribeiro Amaral et al., 2023a), species in a region (Moraga, 2021b), and cells in a tissue (González and Moraga, 2023a). The **spatstat** package (Baddeley et al., 2022) can be used to work with spatial point patterns. The package includes a number of functions that allow us to conduct spatial analysis, such as assessing the randomness of spatial point patterns, and to formulate and fit models to point pattern data.

An example of spatial point pattern is given by the `swedishpines` data from **spatstat**. This pattern represents the locations of 71 trees in a Swedish forest plot of 9.6×10 meters (Figure 17.1).

```
library(spatstat)
data(swedishpines)
X <- swedishpines
plot(X)
axis(1)
axis(2)
summary(X)
```

To get an impression of the spatial point pattern, we can calculate the intensity of events, which indicates the mean number of events per unit area. The `density()` function of the **spatstat** package can be used to compute a kernel smoothed intensity function from a point pattern. This function has an argument called `kernel` that indicates the type of kernel (Gaussian by default), and an argument called `sigma` which refers to the smoothing bandwidth, the standard deviation of the smoothing kernel.

In the `swedishpines` data, the coordinates of the point pattern are expressed in decimeters (0.1 meter). Here, we use `density()` with `sigma = 10` so the smoothing bandwidth is 10 decimeters or 1 meter. Figure 17.1 shows the estimated intensity. We observe that the intensity varies across the region, and the average intensity is equal to 0.0074 trees per square decimeter, that is, 0.74 trees per square meter.

```
# density() calls density.ppp() if the argument is a ppp object
den <- density(x = X, sigma = 10)
summary(den)
plot(den, main = "Intensity")
contour(den, add = TRUE) # contour plot


Planar point pattern: 71 points
Average intensity 0.007396 points per square unit
(one unit = 0.1 metres)

Coordinates are integers i.e. rounded to the nearest unit

Window: rectangle = [0, 96] x [0, 100] units
Window area = 9600 square units
Unit of length: 0.1 metres

real-valued pixel image
128 x 128 pixel array (ny, nx)
enclosing rectangle: [0, 96] x [0, 100] units
dimensions of each pixel: 0.75 x 0.7812 units
Pixel values
    range = [0.001842, 0.01569]
    integral = 71.02
    mean = 0.007398
```

17.1   Examples

The **spatstat** package contains a number of examples of spatial point patterns. Here, we describe some of the data included in **spatstat**, and this document[1] provides an overview of all the data included in the package.

---

[1] https://cran.r-project.org/web/packages/spatstat/vignettes/datasets.pdf

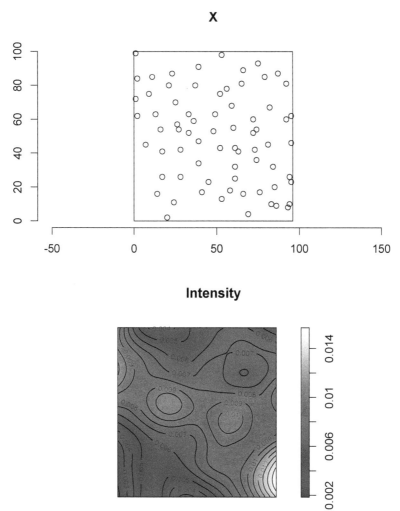

FIGURE 17.1: Locations (top) and intensity (bottom) of 71 trees in a Swedish forest plot.

Japanese pines

The `japanesepines` data from **spatstat** represents locations of 65 saplings of Japanese pine in a 5.7 × 5.7 square meter sampling region in a natural stand (Figure 17.2). An interesting question when analyzing this data could be whether the spacing between saplings is greater than would be expected for a random pattern (which could indicate competition for resources).

```
library(spatstat)
japanesepines
```

```
Planar point pattern: 65 points
window: rectangle = [0, 1] x [0, 1] units (one unit =
5.7 metres)
```

```
plot(japanesepines)
axis(1)
axis(2)
```

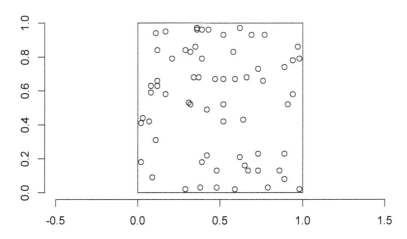

FIGURE 17.2: Locations of 65 saplings of Japanese pine in a natural stand.

Trees in a forest

Spatial point patterns can also have an associated value, and these are called marked point patterns. An example of marked point pattern is given by the longleaf data from **spatstat** which contains locations of 584 trees in a forest of longleaf pine trees in Georgia, USA, along with their diameter at breast height (dbh), a convenient surrogate measure of size and age (Figure 17.3). Here, it would be interesting to understand the spatial variation in the density and age of trees.

```
longleaf
```

```
Marked planar point pattern: 584 points
marks are numeric, of storage type  'double'
window: rectangle = [0, 200] x [0, 200] metres
```

```
plot(longleaf)
axis(1)
axis(2)
```

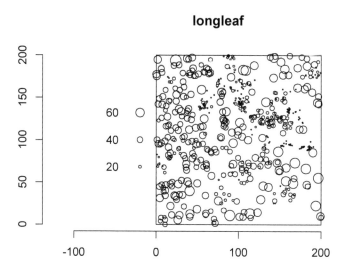

FIGURE 17.3: Locations and diameters of 584 trees in a forest of longleaf pine trees in Georgia, USA.

Castilla-La Mancha forest fires

The clmfires data contains the locations and information of forest fires in the Castilla-La Mancha region of Spain between 1998 and 2007. Figure 17.4 shows the fire locations and four marks with information about each fire, namely, the cause of fire (cause), the total area burned in hectares (burnt.area), the date of fire (date), and the number of days elapsed since 1 January 1998 (julian.date). The main question when analyzing this data could be to understand the spatio-temporal variability of forest fires and potential risk factors.

17 Spatial point patterns

```
clmfires
```

```
Marked planar point pattern: 8488 points
Mark variables: cause, burnt.area, date, julian.date
window: polygonal boundary
enclosing rectangle: [4.1, 391.4] x [18.6, 385.2]
kilometres
```

```
plot(clmfires)
```

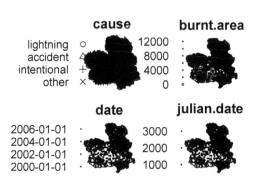

FIGURE 17.4: Locations and information of forest fires in Castilla-La Mancha, Spain.

Hamster tumor data

The `hamster` data provides the centers of the nuclei of certain cells in a section of tissue from a laboratory-induced lymphoma in the kidney of a hamster (Figure 17.5). The nuclei are classified as either "pyknotic" (corresponding to dying cells) or "dividing" (corresponding to cells in the act of dividing). The background void is occupied by unrecorded, interphase cells in relatively large numbers. Using this data, we could investigate how different types of cells interact, and what is the relationship between the degree of cells interaction and cancer stage and survival.

```
hamster
```

Marked planar point pattern: 303 points
Multitype, with levels = dividing, pyknotic
window: rectangle = [0, 1] x [0, 1] units (one unit =
250 microns)

plot(hamster)

hamster

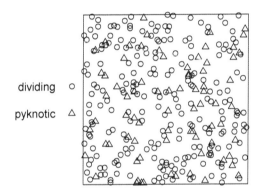

dividing ○

pyknotic △

FIGURE 17.5: Cells in a tissue from a lymphoma in the kidney of a hamster.

Chorley-Ribble data

The `chorley` data gives the addresses of 58 larynx cancer cases and 978 lung cancer cases, recorded in the Chorley and South Ribble Health Authority of Lancashire, England, between 1974 and 1983. Figure 17.6 shows the locations of the case addresses, as well as the location of a disused industrial incinerator. After allowing for spatial variation in the density of the susceptible population, we could assess the evidence for an increase in the incidence of larynx cancer near the incinerator. Here, the lung cancer cases could serve as a surrogate for the spatially varying population density.

chorley

Marked planar point pattern: 1036 points
Multitype, with levels = larynx, lung
window: polygonal boundary
enclosing rectangle: [343.4, 366.4] x [410.4, 431.8]
km

```
plot(chorley)
points(chorley.extra$incin, pch = 10, cex = 2, col = "blue")
```

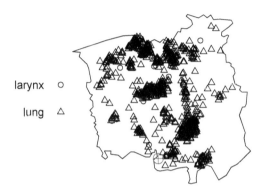

FIGURE 17.6: Locations of larynx and lung cancer cases, and the location
of a disused industrial incinerator in a region of Lancashire, England.

18

The **spatstat** package

The **spatstat** package (Baddeley et al., 2022) can be used for statistical analysis of spatial point patterns. In **spatstat**, spatial point patterns are represented with objects of class ppp that contain the event locations with possibly associated marks, and the observation window where the events occur. Here, we show how to use **spatstat** to create ppp objects representing spatial point patterns, and how to transform ppp objects to sf objects to be able to work with the data using packages such as **sf**.

18.1 Creating spatial point patterns

In **spatstat**, spatial point patterns are represented as objects of class ppp (planar point pattern). To create a ppp object, we use the ppp() function passing the vectors x and y with the event coordinates, and the observation window which is of class owin.

For example, here we create a spatial point pattern of 100 randomly generated points in the region $[0, 1] \times [0, 2]$. First, we use the owin() function to create an object of class owin with the observation window $[0, 1] \times [0, 2]$ passing the ranges of the horizontal and vertical axes.

```
library(spatstat)
win <- owin(xrange = c(0, 1), yrange = c(0, 2))
# plot(win)
```

Then, we simulate 100 random points in the observation window $[0, 1] \times [0, 2]$.

```
x <- runif(100, 0, 1)
y <- runif(100, 0, 2)
```

Finally, we create the ppp object with the ppp() function passing the x and y coordinates of the events, and the owin object with the window.

```
X <- ppp(x = x, y = y, window = win)
X
```

```
Planar point pattern: 100 points
window: rectangle = [0, 1] x [0, 2] units
```

Figure 18.1 shows the plot of the **ppp** object.

```
plot(X)
axis(1)
axis(2)
```

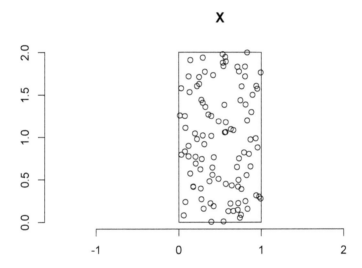

FIGURE 18.1: Point pattern of 100 independent uniform random points generated in $[0, 1] \times [0, 2]$.

We can extract the observation window of the point pattern X with `Window()`.

```
Window(X)
```

```
window: rectangle = [0, 1] x [0, 2] units
```

An alternative way of simulating a spatial point pattern of independent uniform random points in a given region is by using the `runifpoint()` function. The arguments of `runifpoint()` include the number n of points and the window of class **owin** where the point pattern is simulated. For example, the previous pattern could have been generated using X <- runifpoint(n = 100, win = win).

Marks denoting associated information of events can be set with marks() or %mark%. For example, Figure 18.2 shows the previous point pattern X where we add a mark with a numeric value to each of the events.

```
marks(X) <- 1:npoints(X)
X <- X %mark% 1:npoints(X)
plot(X)
```

X

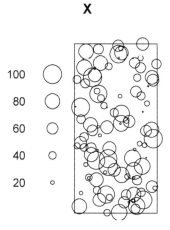

FIGURE 18.2: Point pattern of 100 independent uniform random points generated in $[0, 1] \times [0, 2]$ with numeric marks.

Note that the definition of an observation window that represents the study region of the point pattern needs to be carefully specified, as it affects the visualization and analysis of the data, and possibly the conclusions obtained. For example, if the previous point pattern was thought to be observed in the window $[0, 5] \times [0, 5]$, the data would appear as a cluster in the bottom left corner of the window instead of randomly in $[0, 1] \times [0, 2]$. Note that this situation, depicted in Figure 18.3, would change the interpretation of the data.

```
win2 <- owin(xrange = c(0, 5), yrange = c(0, 5))
X2 <- ppp(x = x, y = y, window = win2)
plot(X2)
axis(1)
axis(2)
```

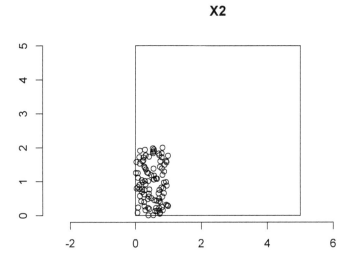

FIGURE 18.3: Point pattern of 100 independent uniform random points generated in $[0,1] \times [0,2]$ plotted in $[0,5] \times [0,5]$.

The `inside.owin()` function can be used to test whether a set of points lie within a particular observation window. For example, we can identify the points in the point pattern X that are inside the unit square by passing the points and the observation window as follows:

```
win <- owin() # unit square observation window
marks(X) <- inside.owin(X, w = win)
plot(X)
axis(1)
axis(2)
```

18.2 Converting between ppp and sf objects

From ppp to sf

In some situations, we may be interested in transforming an object of class `ppp` to an object of class `sf` to be able to manipulate and visualize the data using other packages such as `sf`. Here, we show how to use the `st_as_sf()` function from `sf` to transform the `longleaf` data from **sptatstat** which contains the locations and sizes of long leaf pine trees, from `ppp` to `sf` class.

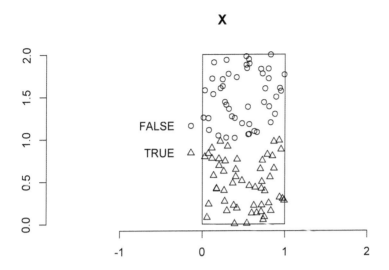

FIGURE 18.4: Point pattern of 100 independent uniform random points generated in $[0, 1] \times [0, 2]$ with labels that indicate whether points lie inside or outside the unit square.

First, we create a data frame containing the coordinates and the marks of the **ppp** object. Then, we create a **sf** object by using the **st_as_sf()** function of **sf** passing the data frame and specifying the name of the columns that contain the event coordinates in argument **coords**. The window of the point pattern can also be converted to **sf** with **st_as_sf(Window(X))**.

```
library(sf)
# ppp object
X <- longleaf
# create data frame with coordinates and marks
ddf <- data.frame(x = X$x, y = X$y, m = marks(X))
# create sf object with data frame and name of coordinates
d <- st_as_sf(ddf, coords = c("x", "y"))
```

From **sf** to **ppp**

We can also convert a **sf** object to a **ppp** object with **as.ppp()** by providing the point coordinates and the observation window. For example, here we obtain the coordinates of the **sf** object in matrix form with **st_coordinates()**, and consider the observation window as the bounding box of the data which can be obtained with `st_bbox()`. Marks of the points can be set with `marks()` or `%mark%`. Figure 18.5 shows the plot of the `ppp` object obtained.

```
X <- as.ppp(st_coordinates(d), st_bbox(d))
marks(X) <- d$m # alternatively we can use X <- X %mark% d$m
plot(X)
```

In case we wish to use a polygon of class `sf` as observation window, we can use `as.owin()` to transform the `sf` object to an `owin` object. Note that the `sf` object needs to be in a projected coordinate reference system. Here, we show an example on how to create a spatial point pattern with the boundary of Brazil as observation window. First, we obtain the Brazil map with the **rnaturalearth** package. Then, we use `st_transform()` to transform the map to projection EPSG 29172 (UTM zone 22N). Finally, we use `runifpoint()` to generate 100 independent uniform random points within the observation window (Figure 18.5).

```
library(rnaturalearth)
map <- ne_countries(type = "countries", country = "Brazil",
                    scale = "medium", returnclass = "sf")
map <- st_transform(map, crs = "EPSG:29172")
win <- as.owin(map)
X <- runifpoint(100, win = win)
plot(X)
```

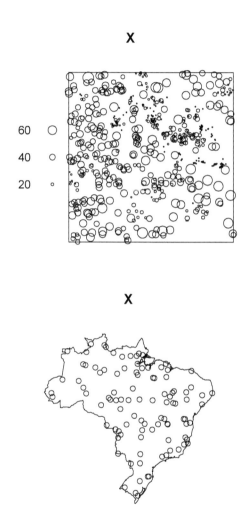

FIGURE 18.5: Spatial point pattern converted from **sf** to **ppp** considering the observation window as the bounding box of the data (top) and as a polygon (bottom).

19

Spatial point processes and simulation

19.1 Spatial point processes

A stochastic process is a collection of random variables X_i where i belongs to an indexing set I, and each X_i takes values in a sample space. A spatial point process $\{X_1, X_2, \ldots, X_{N(A)}\}$ is a stochastic process taking values in $A \subset \mathbb{R}^2$. A realization of a spatial point process is referred to as spatial point pattern and consists of a countable set of points $\{x_1, x_2, \ldots, x_{N(A)}\}$ in the plane. We write $N(A)$ for the number of events in a planar region A, $N(A) = \#(x_i \in A)$. Note that $N(A)$ is a random variable, and therefore, different realizations of the spatial point process may result in both different numbers and locations of events. We commonly refer to the points in the point pattern as events, to differentiate them from arbitrary points in the plane.

Let dx and dy denote small regions containing the points x and y, respectively. The (first-order) intensity function of a spatial point process is defined as

$$\lambda(x) = \lim_{|dx| \to 0} \frac{E[N(dx)]}{|dx|}.$$

The second-order intensity function is

$$\lambda_2(x, y) = \lim_{|dx| \to 0, |dy| \to 0} \frac{E[N(dx)N(dy)]}{|dx||dy|}.$$

The covariance density is expressed as

$$\gamma(x, y) = \lambda_2(x, y) - \lambda(x)\lambda(y).$$

A spatial point process is stationary and isotropic if its statistical properties do not change under translation and rotation, respectively. More specifically, for any integer k and regions $\{A_i : i = 1, \ldots, k\}$, a process is stationary if the joint distribution of $N(A_1), \ldots, N(A_k)$ is invariant to translation by an arbitrary amount x. The process is isotropic if the joint distribution of $N(A_1), \ldots, N(A_k)$ is invariant to rotation through an arbitrary angle θ. That is, if there are no directional effects. Moreover, the process is orderly if there can be no co-located observations so $\lim_{|A|\to 0} P(N(A) > 1) = 0$.

Given a stationary and isotropic spatial point process, the intensity function is a constant equal to the expected number of events per unit area:

$$\lambda(\boldsymbol{x}) = \lambda = \frac{E[N(A)]}{|A|}.$$

Thus, the second-moment intensity reduces to a function of distance:

$$\lambda_2(\boldsymbol{x}, \boldsymbol{y}) = \lambda_2(||\boldsymbol{x} - \boldsymbol{y}||) = \lambda_2(h),$$

where $h = ||\boldsymbol{x} - \boldsymbol{y}||$ is the distance between $\boldsymbol{x}$ and $\boldsymbol{y}$. Moreover, the covariance density is expressed as

$$\gamma(\boldsymbol{x}, \boldsymbol{y}) = \gamma(h) = \lambda_2(h) - \lambda^2.$$

19.2   Poisson processes

Let $\lambda(\cdot)$ be a non-negative valued function, called intensity function of the spatial point process. A Poisson process is characterized by the following properties:

1. The number of events in any region A, $N(A)$, follows a Poisson distribution with mean

$$\mu(A) = \int_A \lambda(\boldsymbol{x}) d\boldsymbol{x}.$$

Thus, $P(N(A) = n) = exp(-\mu(A)) \frac{\mu(A)^n}{n!}$.

2. Given $N(A) = n$, the locations of the n events in A form an independent random sample from the distribution on A with probability density function proportional to the intensity $\lambda(\cdot)$.

Poisson processes can be classified as homogeneous and inhomogeneous Poisson processes. In homogeneous Poisson processes, the intensity is constant ($\lambda(\boldsymbol{x}) = \lambda$, $\forall \boldsymbol{x}$), whereas in inhomogeneous Poisson processes, the intensity varies in space. Homogeneous Poisson processes are also referred to as complete spatial randomness (CSR). In CSR processes, the expected intensity of points is constant across any region, and there is no interaction between the points.

19.3   Simulating Poisson point patterns with rpoispp()

Simulating point patterns is useful when we want to test theoretical properties of the processes and compare them with the data we analyze. Here, we show how to use the rpoispp() function of the **spatstat** package to simulate spatial point patterns from homogeneous and inhomogeneous Poisson processes. The rpoispp() function accepts an argument lambda that indicates the intensity of the Poisson process. lambda can be a single positive number, function, or pixel image. The argument win of rpoispp() is the observation window in which to simulate the pattern. win can be an object of class owin or an object acceptable to as.owin().

19.3.1   Homogeneous point process

Here, we generate a spatial point pattern from a homogeneous Poisson process with intensity $\lambda = 100$ in the window $A = [0,1] \times [0,2]$ which has area $|A| = 2$. To do that, we use the rpoispp() function setting lambda = 100 and win to the observation window owin(xrange = c(0, 1), yrange = c(0, 2)). The generated point pattern is shown in Figure 19.1.

```
library(spatstat)
phom <- rpoispp(lambda = 100,
                win = owin(xrange = c(0, 1), yrange = c(0, 2)))
phom$n
```

```
[1] 205
```

```
plot(phom, main = "Homogeneous")
```

Note that the difference between patterns generated with the rpoispp() and runifpoint() functions. rpoispp() generates a point pattern using an homogeneous Poisson process. In a homogeneous Poisson process, the generated number of points in a window A follows a Poisson distribution with mean $\int_A \lambda(x)dx = \lambda \times |A|$ (expected number of points per unit area $\times$ area of the window), and the points are independent randomly distributed over the window. In our example, the number of points generated is equal to phom$n = 177$. This number has been generated from a Poisson distribution with mean $\int_A \lambda(x)dx = \lambda \times |A| = 100 \times 2 = 200$. On the other hand, runifpoint() generates independent random points in the window conditioning on the total number of points equal to 200.

Homogeneous

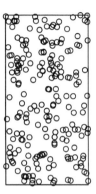

FIGURE 19.1: Point pattern generated from a homogeneous Poisson process.

```
punif <- runifpoint(n = 200,
         win = owin(xrange = c(0, 1), yrange = c(0, 2)))
punif$n
```

```
[1] 200
```

19.3.2   Inhomogeneous point process

We can also use `rpoispp()` to generate a spatial point pattern from an inhomogeneous Poisson process. Here, we consider an intensity function $\lambda(x, y) = 10 + 100 \times x + 200 \times y$ and an observation window equal to $[0, 1] \times [0, 2]$. We write a function to evaluate the intensity function on a fine grid, and we visualize it to see how the intensity varies in space (Figure 19.2).

```
# intensity function
lambda <- function(x){return(10 + 100 * x[1] + 200 * x[2])}

# grid
xseq <- seq(0, 1, length.out = 50)
yseq <- seq(0, 2, length.out = 100)
grid <- expand.grid(xseq, yseq)

# evaluation of the function on a grid
```

```
z <- apply(grid, 1, lambda)

# plot
library(fields)
zmat <- matrix(z, 50, 100)
fields::image.plot(xseq, yseq, zmat, xlab = "x", ylab = "y",
                   main = "lambda(x, y)", asp = 1)
```

In point patterns generated from an inhomogeneous Poisson process, the number of events in any region follows a Poisson distribution with mean equal to the integral of the intensity over the region. Then, the location of each point is independently distributed in the region with probability density proportional to the intensity. In our example, the number of points is generated from a Poisson distribution with mean $\int_{[0,1]\times[0,2]} \lambda(x,y)dxdy = 520$. Figure 19.2 depicts the generated point pattern. We observe a higher number of points located in the regions where the intensity is higher.

```
fnintensity <- function(x, y){return(10 + 100 * x + 200 * y)}
pinhom <- rpoispp(lambda = fnintensity,
                  win = owin(xrange = c(0, 1), yrange = c(0, 2)))
pinhom$n
plot(pinhom, main = "Inhomogeneous")
```

[1] 483

lambda(x, y)

Inhomogeneous

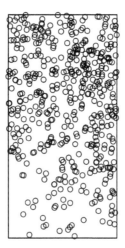

FIGURE 19.2: Intensity of an inhomogeneous Poisson process (top) and point pattern generated from that Poisson process (bottom).

20

Complete spatial randomness

Point processes provide models for point patterns, with complete spatial randomness (CSR) being the simplest theoretical model. CSR assumes that events have an equal likelihood of occurring anywhere within the study area, independent of the locations of other events, which is represented by the homogeneous Poisson process (Diggle, 2014). While most processes deviate from CSR to some degree, CSR remains important in investigations, as it helps differentiate between regular and clustered patterns (Figure 20.1). In a random pattern, the distribution of each point is independent of the distribution of the others, and points neither inhibit nor promote one another. Regular patterns have more spacing between points that in a random pattern, possibly due to mechanisms such as competition preventing close occurrences. Clustered patterns exhibit greater aggregation of points than in a random pattern, likely due to processes such as reproduction with limited dispersal or underlying spatial heterogeneity.

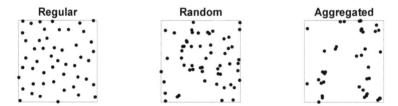

FIGURE 20.1: Examples of regular, random, and aggregated point patterns.

20.1 Testing CSR with the quadrat method

Given a point pattern, the first question is often whether there is any evidence to allow rejection of the null hypothesis of complete spatial randomness (CSR). A simple method used to test CSR is the χ^2 test based on quadrat counts.

The quadrat method partitions the study region into r rows and c columns, which define $m = rc$ non-overlapping subregions or quadrats of equal area. This method relies on the fact that, under CSR, the expected number of observations within any region of equal size is the same. Let n be the number of observed points, m the number of quadrats of equal size, and n_i the number of points in quadrat i. The expected number of points in each quadrat is $n^* = n/m$. The test statistic is calculated as

$$X^2 = \sum_{i=1}^{m} \frac{(\text{observed}_i - \text{expected})^2}{\text{expected}} = \sum_{i=1}^{m} \frac{(n_i - n^*)^2}{n^*}.$$

It can be shown that under CSR, the statistic X^2 has a χ^2_{m-1} distribution. The quadrat method assesses whether CSR is reasonable by comparing the observed value of the X^2 statistic to the χ^2_{m-1} distribution, where m is the number of quadrats. Significance can also be assessed by using Monte Carlo, which involves generating multiple patterns under the null hypothesis, and calculating the X^2 statistic for each of them. The p-value is then determined by comparing the X^2 statistic for the observed point pattern with the X^2 values obtained from the simulations.

Note that the quadrat method's results may depend on the quadrat's configuration. In addition, the method tests CSR for the whole point pattern and cannot distinguish different patterns locally. Later, we will see how the K-function can be used to test CSR at a set of distances overcoming these limitations.

20.2   Example

In this example, we show how to use the `quadrat.test()` function of **spatstat** (Baddeley et al., 2022) to test CSR for a given point pattern based on quadrat counts. We use the **swedishpines** data of **spatstat** which represents the positions of 71 trees in a Swedish forest plot.

```
library(spatstat)
data(swedishpines)
X <- swedishpines
X
```

```
Planar point pattern: 71 points
window: rectangle = [0, 96] x [0, 100] units (one unit
= 0.1 metres)
```

The `quadratcount()` function of **spatstat** divides the window containing the point pattern into nx × ny grid rectangular tiles or quadrats of equal size, and counts the number of points in each quadrat. If the window is

not a rectangle, quadrats are intersected with the window. The arguments
of the `quadratcount()` function include X, the point pattern of class `ppp`,
and `nx` and `ny` which denote the numbers of rectangular quadrats in the x
and y directions (or alternatively `xbreaks` and `ybreaks` giving the x and y
coordinates, respectively, of the quadrats). Figure 20.2 shows the number of
points in each of the quadrats of a 4 × 3 division of the observation window
created with the `quadratcount()` function.

```
Q <- quadratcount(X, nx = 4, ny = 3)
Q
```

	x			
y	[0,24)	[24,48)	[48,72)	[72,96]
[66.7,100]	7	3	6	5
[33.3,66.7)	5	9	7	7
[0,33.3)	4	3	6	9

```
plot(X)
axis(1)
axis(2)
plot(Q, add = TRUE, cex = 2)
```

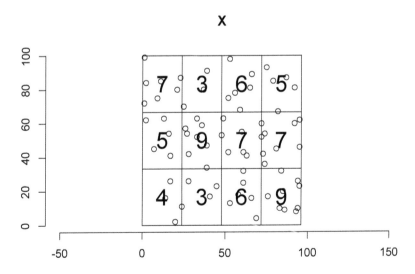

FIGURE 20.2: Number of points in each of the quadrats of a 4 × 3 division
of the observation window.

The function `quadrat.test()` performs a test of CSR for a given point pattern. The first argument can be a point pattern of class `ppp` or the results of applying `quadratcount()` to a point pattern. The alternative hypothesis is specified in argument `alternative` and can take the following values:

- `alternative = "two.sided"` tests H_0: CSR vs. H_1: no CSR (regular or clustered),
- `alternative = "regular"` tests H_0: CSR or clustered vs. H_1: regular,
- `alternative = "clustered"` tests H_0: CSR or regular vs. H_1: clustered.

Here, we use `quadrat.test()` with the default value `alternative = "two.sided"` to test H_0: CSR vs. H_1: no CSR (regular or clustered). By default, `quadrat.test()` assesses significance by comparing the observed test statistic with the chi-squared distribution (`method = "Chisq"`) but can also perform Monte Carlo based tests with `method = "MonteCarlo"`.

```
quadrat.test(Q)
```

```
        Chi-squared test of CSR using quadrat counts

data:
X2 = 7.6, df = 11, p-value = 0.5
alternative hypothesis: two.sided

Quadrats: 4 by 3 grid of tiles
```

The observed value of the test statistic is

$$
\begin{aligned}
X^2 &= \sum_{i=1}^{m} \frac{(\text{observed}_i - \text{expected})^2}{\text{expected}} = \sum_{i=1}^{m} \frac{(n_i - n^*)^2}{n^*} = \\
&= \frac{(7 - n^*)^2 + (3 - n^*)^2 + \cdots + (9 - n^*)^2}{n^*} = 7.59
\end{aligned}
$$

where $m = 12$ is the number of regions of equal size, n_i is the number of points in quadrat i, $i = 1, \ldots, m$, $n = 71$ is the number of observed events, and $n^* = n/m = 71/12 = 5.92$ is the expected number of points in each quadrat.

```
ns <- 71/12
ni <- c(7, 3, 6, 5, 5, 9, 7, 7, 4, 3, 6, 9)
(chi2 <- sum((ni - ns)^2/ns))
```

```
[1] 7.592
```

Under the null hypothesis of CSR, the test statistic has a chi-squared distribution with $m - 1 = 12 - 1$ degrees of freedom. That is, $X^2 \sim \chi^2_{11}$. The p-value is calculated as the probability of obtaining a test statistic as extreme or more extreme than the one observed in the direction of the alternative hypothesis, assuming the null hypothesis is true. If the pattern is regular or clustered, the observed X^2 statistic will be near 0 or large. Therefore, the p-value is calculated as 2 times the minimum of the area to the left of the observed X^2 and the area to the right of the observed X^2:

```
2*min(pchisq(chi2, 11), 1-pchisq(chi2, 11))
```

```
[1] 0.5013
```

The p-value obtained is greater than the level of significance 0.05. So we fail to reject the null hypothesis and conclude there is no evidence against CSR.

20.3 Alternative hypothesis

Here, we present some examples that use `quadrat.test()` to test hypotheses with each of the possible values of the argument `alternative`. Specifically, we use

- `alternative = "two.sided"` to test H_0: CSR vs. H_1: no CSR (regular or clustered),
- `alternative = "regular"` to test H_0: CSR or clustered vs. H_1: regular, and
- `alternative = "clustered"` to test H_0: CSR or regular vs. H_1: clustered.

In each of the examples, the p-value is calculated as the probability of obtaining a test statistic as extreme or more extreme than the one observed in the direction of the alternative hypothesis, assuming the null hypothesis is true. If the calculated p-value is smaller than the significance level α, we would reject the null hypothesis. Otherwise, we would fail to reject the null hypothesis.

To test H_0: CSR or clustered vs. H_1: regular, we use `alternative = "regular"`. If the pattern is regular, the observed X^2 statistic will be near 0. Then, the p-value is calculated as the area to the left of the observed X^2 (Figure 20.3).

```
quadrat.test(Q, alternative = "regular")
```

```
        Chi-squared test of CSR using quadrat counts
```

```
data:
X2 = 7.6, df = 11, p-value = 0.3
alternative hypothesis: regular
```

```
Quadrats: 4 by 3 grid of tiles
```

```
# p-value is area to the left of chi2
pchisq(chi2, 11)
```

```
[1] 0.2506
```

To test H_0: CSR or regular vs. H_1: clustered, we use `alternative = "clustered"`. If the pattern is clustered, the observed X^2 statistic will be large, and the p-value is calculated as the area to the right of the observed X^2 (Figure 20.3).

```
quadrat.test(Q, alternative = "clustered")
```

```
    Chi-squared test of CSR using quadrat counts
```

```
data:
X2 = 7.6, df = 11, p-value = 0.7
alternative hypothesis: clustered
```

```
Quadrats: 4 by 3 grid of tiles
```

```
# p-value is area to the right of chi2
1-pchisq(chi2, 11)
```

```
[1] 0.7494
```

Finally, we test H_0: CSR vs. H_1: no CSR (regular or clustered) using the default value `alternative = "two.sided"`. If the pattern is regular or clustered, the observed X^2 statistic will be near 0 or large. The p-value is calculated as 2 times the minimum of the area to the left of the observed X^2 and the area to the right of the observed X^2.

```
quadrat.test(Q, alternative = "two.sided") # default
```

```
    Chi-squared test of CSR using quadrat counts
```

```
data:
X2 = 7.6, df = 11, p-value = 0.5
alternative hypothesis: two.sided

Quadrats: 4 by 3 grid of tiles
```

```
# p-value is 2 times the minimum of
# area to the left of observed chi-squared and
# area to the right of observed chi-squared
2*min(pchisq(chi2, 11), 1-pchisq(chi2, 11))
```

[1] 0.5013

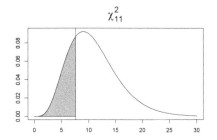 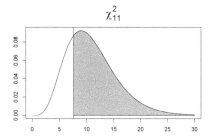

FIGURE 20.3: Area under the χ^2_{11} density curve corresponding to the p-value for alternative hypothesis regular (left) and clustered (right).

21

Intensity estimation

The intensity function of a spatial point process $\{X_1, X_2, \ldots, X_{N(A)}\}$ in a planar window $A \subset \mathbb{R}^2$ is defined as

$$\lambda(\boldsymbol{x}) = \lim_{|d\boldsymbol{x}| \to 0} \frac{E[N(d\boldsymbol{x})]}{|d\boldsymbol{x}|},$$

where $d\boldsymbol{x}$ is a small region containing the point $\boldsymbol{x}$. Given a stationary and isotropic spatial point process, the intensity function is constant equal to the expected number of events per unit area:

$$\lambda(\boldsymbol{x}) = \lambda = \frac{E[N(A)]}{|A|}.$$

Thus, for an observed spatial point pattern of n events observed in a region A, the intensity can be estimated as the observed number of events per unit area:

$$\hat{\lambda} = \frac{n}{|A|}$$

For non-stationary processes, a common method to estimate the spatially varying intensity function involves kernel density estimation (Silverman, 1986; González and Moraga, 2023b). Usually, kernel estimation methods focus on estimating the probability density function $f(\cdot)$ rather than the intensity function $\lambda(\cdot)$. The density function defines the probability of observing an event at a location $\boldsymbol{x}$ and integrates to one across the area of study. In contrast, the intensity function provides the number of events expected per unit area at location $\boldsymbol{x}$ and integrates to the overall mean number of events per unit area. As a result, the density and intensity functions are proportional:

$$\lambda(\boldsymbol{x}) = f(\boldsymbol{x}) \int_A \lambda(\boldsymbol{u}) d\boldsymbol{u},$$

where $\int_A \lambda(\boldsymbol{u}) d\boldsymbol{u}$ is the expected number of events in A. Then, the relative spatial pattern in densities and intensities are the same.

Kernel estimators of the density function $f(\cdot)$ and the intensity function $\lambda(\cdot)$ at the location $\boldsymbol{x}$ based on the observed events $\{\boldsymbol{x}_1, \ldots, \boldsymbol{x}_n\}$ take the form

$$\hat{f}(\boldsymbol{x}) = \frac{1}{n} \sum_{i=1}^{n} \frac{1}{h^2} K\left(\frac{\boldsymbol{x} - \boldsymbol{x}_i}{h}\right) \quad \text{and}$$

$$\hat{\lambda}(\boldsymbol{x}) = \sum_{i=1}^{n} \frac{1}{h^2} K\left(\frac{\boldsymbol{x} - \boldsymbol{x}_i}{h}\right),$$

where $K(\cdot)$ is a symmetric function such that $K(\boldsymbol{x}) \geq 0 \; \forall \boldsymbol{x}$ and $\int_A K(\boldsymbol{x}) d\boldsymbol{x} = 1$ known as kernel, and h is a smoothing parameter known as "bandwidth" (Figure 21.1). Common choices for the kernel include the following functions:

- Gaussian kernel: $K(\boldsymbol{x}) = \frac{1}{\sqrt{2\pi}} exp(-\boldsymbol{x}^2/2)$

- Epanechnikov kernel: $K(\boldsymbol{x}) = \frac{3}{4}(1 - \boldsymbol{x}^2) I(|\boldsymbol{x}| < 1)$

- Quartic kernel: $K(\boldsymbol{x}) = \frac{15}{16}(1 - \boldsymbol{x}^2)^2 I(|\boldsymbol{x}| < 1)$

- Uniform kernel: $K(\boldsymbol{x}) = \frac{1}{2} I(|\boldsymbol{x}| < 1)$

Edge effects tend to distort the kernel estimates close to the boundary of the region, since events near the boundary have fewer local neighbors than events in the interior. One way to deal with this problem is to modify the kernel estimate by dividing by the following edge-correction term:

$$p_h(\boldsymbol{x}) = \int_A h^{-2} K\left(\frac{\boldsymbol{x} - \boldsymbol{u}}{h}\right) d\boldsymbol{u},$$

which represents the volume under the scaled kernel centered on $\boldsymbol{x}$ which lies inside the study region (Gatrell et al., 1996).

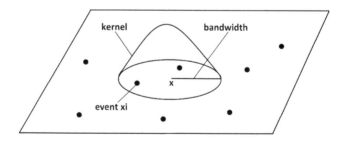

FIGURE 21.1: Kernel estimation representation.

21.1 Intensity

The `density()` function of **spatstat** can be used to obtain a kernel estimate of the intensity of a point pattern. The arguments of `density()` can be seen by typing `?density.ppp`. These include the type of kernel (`kernel`) and the smoothing bandwidth (`sigma`). By default, `density()` uses a Gaussian kernel and a bandwidth determined by a simple rule of thumb that depends only on the size of the window. Here, we use the `density()` function to estimate the intensity of the point pattern of tree locations that is in the `japanesepines` data from **spatstat** (Figure 21.2).

```
library(spatstat)
X <- japanesepines
plot(X)
axis(1)
axis(2)
```

The `density()` function returns an intensity estimate object of class `im` that can be plotted (Figure 21.2). The kernel bandwidth used in `density()` can be extracted with the `sigma` attribute of the returned intensity object.

```
lambdahat <- density(X)
attr(lambdahat, "sigma")
```

```
[1] 0.125
```

Note that although the type kernel weakly influences the estimates, the bandwidth can have a big impact. For example, Figure 21.2 shows intensity estimates of the tree point pattern using different values of bandwidths. We observe that small values of the bandwidth result in estimated intensities that are too spiky, whereas large values provide smoother surfaces that may ignore local characteristics of the intensities.

```
plot(lambdahat, main = "Default bandwidth")
plot(density(X, sigma = 0.05))
plot(density(X, sigma = 0.1))
plot(density(X, sigma = 1))
```

In practice, we may conduct exploratory analyses considering a number of possible bandwidth values to select, somewhat subjectively, an appropriate bandwidth. Other criteria such as cross-validation as implemented in the `bw.diggle()` and `bw.ppl()` functions of **spatstat** can be used to select a

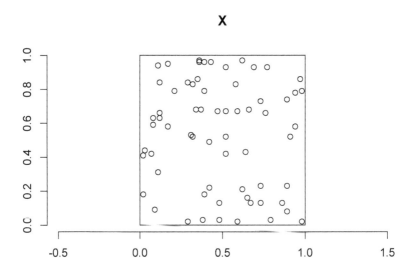

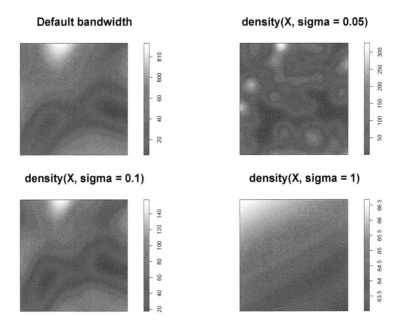

FIGURE 21.2: Top: Trees point pattern. Bottom: Intensity estimates using several values for the kernel bandwidth.

smoothing bandwidth for the kernel estimation of the point process intensity. We can also use adaptative kernel estimators where bandwidths change at each data point of the spatial point pattern (González and Moraga, 2022, 2023c).

21.2 Intensity ratio

In some situations, we may want to compare two point patterns observed in the same region, such as the patterns of individuals with a disease, and set of controls representing at-risk population. We can do that by computing the intensity ratio of the patterns, which allows us to identify spatial patterns and hotspots in the relative risk surface obtained.

Here, we show how to estimate the intensity ratio of two point patterns using the `density()` function of **spatstat**, and we visualize the results with the `image.plot()` function of **fields**. Alternatively, we could estimate the intensity ratio by using the **sparr** package which provides functions to estimate and assess the significance of relative risk surfaces.

We consider the data `pbc` from the **sparr** package which contains 761 cases of primary biliary cirrhosis (PBC) along with 3020 controls representing at-risk population in north-eastern England collected between 1987 and 1994. This data is represented in Figure 21.3. The `pbc` data is a `ppp` object with marks `case` and `control` representing the cases and controls for the PBC data in north-eastern England. We create the `ppp` objects `cases` and `controls` with the events corresponding to each type.

```
library(sparr)
data(pbc)
cases <- unmark(pbc[which(pbc$marks == "case"), ])
plot(cases, main = "cases")
axis(1)
axis(2)
title(xlab = "Easting", ylab = "Northing")
controls <- unmark(pbc[which(pbc$marks == "control"), ])
plot(controls, pch = 3, main = "controls")
axis(1)
axis(2)
title(xlab = "Easting", ylab = "Northing")


library(sparr)
data(pbc)
```

```
cases <- unmark(pbc[which(pbc$marks == "case"), ])
controls <- unmark(pbc[which(pbc$marks == "control"), ])
```

We assume that the point pattern cases is a realization from a Poisson process with intensity $\lambda(x)$, and that the point pattern controls come from a second, independent Poisson process with intensity $\lambda_0(x)$. Then, we can express $\lambda(x)$ as

$$\lambda(x) = \alpha \lambda_0(x) \rho(x).$$

Here, $\rho(x)$ represents the spatial variation in relative risk. α is a factor that adjusts the intensity estimate of the controls to take account that there are more controls than cases. This factor can be estimated as $\hat{\alpha} = $ (number of cases)/(number of controls). An estimate of the density ratio can then be obtained as the ratio of the kernel estimates of the intensity of cases and controls as follows:

$$\hat{\rho}(x) = \frac{\hat{\lambda}(x)}{\hat{\alpha}\hat{\lambda}_0(x)}.$$

Here, we show how to compute and plot the relative risk for the PBC case-control data. First, we obtain a common bandwidth to obtain the kernel estimates of the intensities of cases and controls. We calculate this common bandwidth as the mean of the default bandwidths obtained when using density() to estimate the intensity of cases and controls separately.

```
bwcases <- attr(density(cases), "sigma")
bwcontr <- attr(density(controls), "sigma")
(bw <- (bwcases + bwcontr)/2)
```

```
[1] 11.46
```

Then, we use the selected bandwidth to compute the smoothed intensity estimates for the cases and controls.

```
intcases <- density(cases, sigma = bw)
intcontrols <- density(controls, sigma = bw)
```

We estimate α as the ratio of the number of cases (cases$n = 761) to the number of controls (controls$n = 3020) to account for the fact that there are more controls than cases.

```
(alphahat <- cases$n/controls$n)
```

```
[1] 0.252
```

Then, using the intensity estimates of the cases and controls patterns, the relative risk is estimated as

$$\hat{\rho}(\boldsymbol{x}) = \frac{\hat{\lambda}(\boldsymbol{x})}{\hat{a}\hat{\lambda}_0(\boldsymbol{x})}.$$

Figure 21.3 shows the estimated intensity ratio obtained with the image.plot() function of the **fields** package. Note that to plot the intensity estimate object with image.plot(), we first need to transpose the image values returned by density() since they are stored in transposed form.

```
library(fields)
x <- intcases$xcol
y <- intcases$yrow
rr <- t(intcases$v)/t(alphahat * intcontrols$v)
image.plot(x, y, rr, asp = 1)
```

21.3 Intensity on networks

The **spatstat** package also contains functionality to work with spatial point patterns on linear networks. An example of this type of point pattern is given by the chicago data of **spatstat**. This data contains the locations and type of crimes reported from 25 April to 8 May 2002 in an area of Chicago, Illinois, USA. All crimes occurred on or near a street, and the data provide the coordinates of all streets in the survey area, and their connectivity. Spatial coordinates are expressed in feet (1 foot corresponds to 0.3048 meters). Figure 21.4 shows the crime locations by type and the streets of the study region.

```
library(spatstat)
X <- chicago
head(X$data)
X$domain
plot(X)
```

```
Hyperframe:
      x      y seg      tp    marks
1 639.2 1191   37 1.00000  assault
2 139.7 1135   54 0.36985  assault
3 195.3 1150   53 0.74153  assault
```

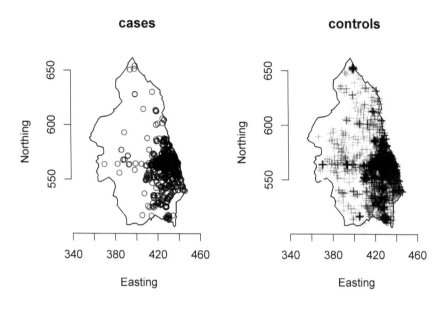

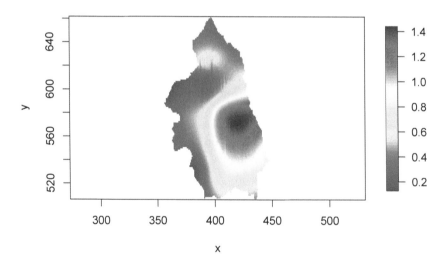

FIGURE 21.3: Cases of primary biliary cirrhosis and controls representing at-risk population (top) and intensity ratio (bottom) in north-eastern England between 1987 and 1994.

```
4 693.5 1125   72 0.02486 assault
5 596.4 1005 122 0.32558 assault
6 684.7 1005 124 0.90250 assault
```

```
Linear network with 338 vertices and 503 lines
Enclosing window: rectangle = [0.4, 1282] x [153.1,
1276.6] feet
```

chicago is an object of class lpp (point pattern on a linear network). lpp objects contain the locations of the points as an ppp object or other object acceptable to as.ppp(), and a linear network of class linnet.

The density.lpp() function of **spatstat** allows us to obtain an estimate of the density of the events on the linear network by applying kernel smoothing. For example, Figure 21.4 shows the intensity of all types of crimes obtained using density.lpp() with a bandwidth of 10.

```
# unmarked point pattern
uX <- unmark(chicago)
# intensity estimate
lambdahat <- density.lpp(uX, sigma = 10)
# plot
plot(lambdahat, main = "Intensity")
```

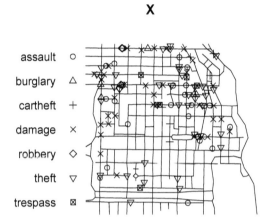

FIGURE 21.4: Crime locations by type (top) and intensity of crime locations (bottom) in an area of Chicago.

22

The K-function

The K-function for a spatial point pattern $\{x_1, \ldots, x_n\}$ observed in a planar window $A \subset \mathbb{R}^2$, is an exploratory tool that can be used to assess the dependence between locations at several distances. The K-function is defined as

$$K(s) = \lambda^{-1} E[\text{number of further events within distance s of an arbitrary event}],$$

where λ is the intensity function of the spatial point process.

For clustered spatial point patterns, each event is likely to be surrounded by further events. Therefore, for small values of the distance s, $K(s)$ will be relatively large. For regular point patterns, each event is likely to be surrounded by empty space. This implies that for small values of s, $K(s)$ will be relatively small.

To determine whether the values of a K-function are relatively large or small, we can compare the K-function for the observed spatial point pattern with the K-function for a homogeneous Poisson process (CSR) that is given by $K(s) = \pi s^2$ as shown below. Thus, for a distance s, $K(s) > \pi s^2$ indicates clustering, and $K(s) < \pi s^2$ suggests inhibition.

Let C be the region denoting the circle of center x_c and radius s, and $|C| = \pi s^2$ the area of the circle. Under complete spatial randomness (CSR), the intensity λ is constant equal to the observed number of events per unit area, and

$$E[\text{number of further events within distance s of an arbitrary event}] =$$
$$= \frac{\text{number of points in } C}{|C|} \times |C| = \lambda \times \pi s^2.$$

Then, under CSR,

$$\lambda K(s) = E[\text{number of further events within distance s of an arbitrary event}] =$$
$$= \lambda \times \pi s^2,$$

which implies $K(s) = \pi s^2$.

22.1   Estimating the K-function

Given a spatial point pattern $\{x_1, \ldots, x_n\}$ in a planar window A, we can construct an estimate of $K(s)$ as follows. First, we define

$$E(s) = E[\text{number of further events within distance s of an arbitrary event}] =$$

$$= \lambda K(s).$$

An estimate of $E(s)$ can be computed as

$$\tilde{E}(s) = \frac{1}{n} \sum_{i=1}^{n} \sum_{j \neq i} I(d_{ij} \leq s),$$

where d_{ij} is the distance between the events x_i and x_j, and $I(\cdot)$ is the indicator function (Figure 22.1).

The estimate $\tilde{E}(s)$ is negatively biased because we do not observe events outside A. This implies the observed counts for events x_i close to the boundary of A may be artificially low. To address this issue, we can introduce weights w_{ij} equal to the reciprocal of the proportion of the circle with center x_i and radius d_{ij} which is contained in A (Figure 22.1). Then, an edge-corrected estimate for $E(s)$ is given by

$$\hat{E}(s) = \frac{1}{n} \sum_{i=1}^{n} \sum_{j \neq i} w_{ij} I(d_{ij} \leq s).$$

The intensity of a spatial point process denotes the expected number of events per unit area. In a homogeneous process, the intensity is constant and can be estimated as $\hat{\lambda} = n/|A|$. Then, since $K(s) = E(s)/\lambda$, the estimate of the K-function can be calculated as

$$\hat{K}(s) = \frac{\hat{E}(s)}{\hat{\lambda}} = \frac{|A|}{n^2} \sum_{i=1}^{n} \sum_{j \neq i} w_{ij} I(d_{ij} \leq s).$$

22.2   The Kest() function

Given a spatial point pattern, the K-function can be estimated using the Kest() function of **spatstat** passing the spatial point pattern X as a **ppp** object or an

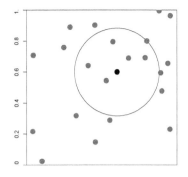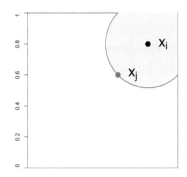

FIGURE 22.1: Left: Point pattern in the unit square region. The black point represents an arbitrary event. The circle encloses the events considered to estimate the K-function at the distance given by the radius of the circle. Right: Gray area represents the inverse of the weight used in the estimation of the K-function using the x_i and x_j events.

object acceptable to `as.ppp()`. The `Kest()` function computes the K-function using three different edge-correction methods, namely, `border`, `isotropic` and `translate`, as well as the theoretical K-function for the homogeneous Poisson process.

Here, we use `Kest()` to estimate the K-function of a simulated spatial point pattern from a homogeneous Poisson process with intensity $\lambda = 100$ in the region $[0,1] \times [0,1]$. Figure 22.2 shows the simulated point pattern with the `rpoispp()` function.

```
library(spatstat)
X <- rpoispp(lambda = 100)
plot(X)
axis(1)
axis(2)
```

Figure 22.3 depicts the K-function calculated for this point pattern using different edge-correction methods together with the theoretical K-function for the homogeneous Poisson process.

```
K <- Kest(X)
plot(K)
```

The L-function is a commonly used function that transforms the K-function corresponding to a homogeneous Poisson process $(K(s) = \pi s^2)$ to a straight line $L(s) = s$ making visual interpretation easier. The L-function is defined as

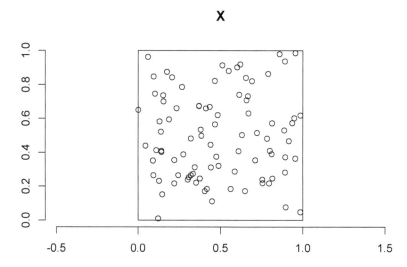

FIGURE 22.2: Simulated point pattern from a homogeneous Poisson process.

$$L(s) = \sqrt{\frac{K(s)}{\pi}}.$$

The `Lest()` function can be used to estimate the L-function of a spatial point pattern. Figure 22.3 shows the L-functions for the simulated point pattern using different edge-correction methods, and the theoretical L-function for the homogeneous Poisson process.

```
L <- Lest(X)
plot(L)
```

22.3   Testing complete spatial randomness

We can use the K-function to test complete spatial randomness (CSR) at a set of distances. Specifically, we can compare the K-function estimate from the data, $\hat{K}(s)$, with the theoretical value of the K-function under CSR, ($K(s) = \pi s^2$). Typically, the estimated K-function does not lie exactly over the line πs^2 representing the theoretical K-function under CSR. Therefore, to better assess CSR, we obtain a confidence region by simulating spatial point

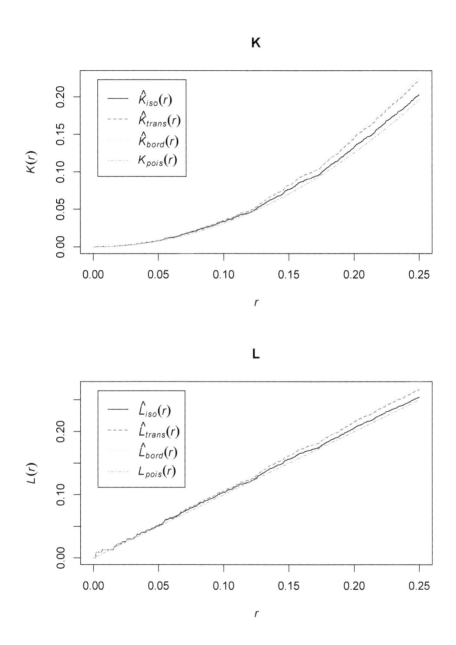

FIGURE 22.3: K-function (top) and L-function (bottom) corresponding to a simulated point pattern from a homogeneous Poisson process.

patterns under CSR. Then, we add the confidence region to the plot of the estimated K-function of the observed point pattern. This plot allows us to assess CSR by comparing the K-function corresponding to the observed data to the envelope for each of the distances. In more detail, we follow this approach:

- Generate a number of spatial point patterns (e.g., $M = 99$, $M = 999$) of the same size as the observed pattern over the study region using a homogeneous Poisson process (CSR).
- For each spatial point pattern, estimate the K-function: $\hat{K}_1(\cdot), \ldots, \hat{K}_M(\cdot)$.
- For each distance s, compute the 95% quantile interval of $\hat{K}_1(s), \ldots, \hat{K}_M(s)$.
- Reject the null hypothesis of CSR if the observed K-function at a given distance is outside the interval.

This approach can be conducted by using the `envelope()` function of **spatstat** which performs simulations and computes envelopes of a summary statistic based on the simulations. Specifically, `envelope()` generates `nsim` simulated point patterns each being a realization of a homogeneous Poisson point process (CSR) with the same intensity as the observed point pattern X. Figure 22.4 shows the envelope obtained for 99 simulations under CSR. The confidence region obtained can be inspected to identify distances for which there is an indication of clustering or inhibition.

```
E <- envelope(X, Kest, nsim = 99)
plot(E)
```

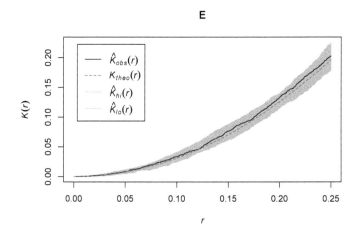

FIGURE 22.4: K-function of a simulated point pattern from a homogeneous Poisson process, together with an envelope corresponding to the K-functions of 99 simulated point patterns under CSR.

23

Point process modeling

23.1 Log-Gaussian Cox processes

Log-Gaussian Cox processes (LGCPs) are typically used to model phenomena that are environmentally driven (Diggle et al., 2013). A LGCP is a Poisson process with a varying intensity, which is itself a stochastic process of the form

$$\Lambda(\boldsymbol{s}) = exp(Z(\boldsymbol{s})),$$

where $Z = \{Z(\boldsymbol{s}) : \boldsymbol{s} \in \mathbb{R}^2\}$ is a Gaussian process. Then, conditional on $\Lambda(\cdot)$, the point process is an inhomogeneous Poisson process with intensity $\Lambda(\cdot)$. This implies that the number of events in any region A is Poisson distributed with mean $\int_A \Lambda(\boldsymbol{s})d\boldsymbol{s}$, and the locations of events are an independent random sample from the distribution on A with probability density proportional to $\Lambda(\cdot)$. A LGCP model can also include spatial explanatory variables providing a flexible approach for describing and predicting a wide range of spatial phenomena.

In this chapter, we assume that we have observed a spatial point pattern of event locations $\{\boldsymbol{x}_i : i = 1, \ldots, n\}$ that has been generated as a realization of a LGCP, and we show how to fit a LGCP model to the data using the INLA and SPDE approaches. Chapter 15 introduced the SPDE approach and described its implementation in the context of model-based geostatistics using an example of air pollution prediction. Here, we describe how to use SPDE to fit a LGCP model to a point pattern of plant species to estimate the intensity of the process.

23.2 Fitting a LGCP

Grid approach

A common method for inference in LGCP models is to approximate the latent Gaussian field by means of a gridding approach (Illian et al., 2012). In this approach, the study region is first discretized into a regular grid of

$n_1 \times n_2 = N$ cells, $\{s_{ij}\}$, $i = 1, \ldots, n_1$, $j = 1, \ldots, n_2$. In the LGCP, the mean number of events in cell s_{ij} is given by the integral of the intensity over the cell, $\Lambda_{ij} = \int_{s_{ij}} exp(\eta(s))ds$, and this integral can be approximated by $\Lambda_{ij} \approx |s_{ij}|exp(\eta_{ij})$, where $|s_{ij}|$ is the area of the cell s_{ij}. Then, conditional on the latent field η_{ij}, the observed number of locations in grid cell s_{ij}, y_{ij}, are independent and Poisson distributed as follows,

$$y_{ij}|\eta_{ij} \sim Poisson(|s_{ij}|exp(\eta_{ij})).$$

Then, the LGCP model can be expressed within the generalized linear mixed model framework. For example, the log-intensity of the Poisson process can be modeled using covariates and random effects as follows:

$$\eta_{ij} = c(s_{ij})\beta + f_s(s_{ij}) + f_u(s_{ij}).$$

Here, $\boldsymbol{\beta} = (\beta_0, \beta_1, \ldots, \beta_p)'$ is the coefficient vector of $\boldsymbol{c}(s_{ij})$ — $(1, c_1(s_{ij}), \ldots, c_p(s_{ij}))$, the vector of the intercept and covariates values at s_{ij}. $f_s()$ is a spatially structured random effect reflecting unexplained variability specified as a second-order two-dimensional conditional autoregressive model on a regular lattice. $f_u()$ is an unstructured random effect reflecting independent variability in the cells. Moraga (2021b) provides an example on how to implement the grid approach to fit a LGCP with INLA using a species distribution modeling study in Costa Rica.

Going off the grid

While the previous approach is a common method for inference in LGCP models, the results obtained depend on the construction of a fine regular grid that cannot be locally refined. An alternative computationally efficient method to perform computational inference on LGCP is presented in Simpson et al. (2016). This method, rather than defining a Gaussian random field over a fine regular grid, proposes a finite-dimensional continuously specified random field of the form

$$Z(\boldsymbol{s}) = \sum_{i=1}^{n} z_i \phi_i(\boldsymbol{s}),$$

where $\boldsymbol{z} = (z_1, \ldots, z_n)'$ is a multivariate Gaussian random vector and $\{\phi_i(\boldsymbol{s})\}_{i=1}^{n}$ is a set of linearly independent deterministic basis functions. Unlike the lattice approach, this method models observations considering its exact location instead of binning them into cells. Thus, the implementation of this method does not need the specification of a regular grid but a triangulated mesh to approximate the Gaussian random field using the SPDE approach. Below, we give an example of species distribution modeling where we fit a

LGCP using this method and INLA and SPDE for fast approximate inference. This approach is also explained in Krainski et al. (2019).

23.3 Species distribution modeling

Species distribution models allow us to understand spatial patterns, and assess the influence of factors on species occurrence. These models are crucial for the development of appropriate strategies that help protect species and the environments where they live. Here, we show how to formulate and fit a LGCP model for *Solanum* plant species in Bolivia using a continuously Gaussian random field with INLA and SPDE. The model allows us to estimate the intensity of the process that generates the locations.

23.3.1 Observed *Solanum* plant species in Bolivia

In this example, we estimate the intensity of *Solanum* plant species in Bolivia from January 2015 to December 2022 which are obtained from the Global Biodiversity Information Facility (GBIF) database with the **spocc** package. We retrieve the data using the `occ()` function specifying the plant species scientific name, data source, dates, and country code. We also specify `has_coords = TRUE` to just return occurrences that have coordinates, and `limit = 1000` to specify the limit of the number of records.

```
library("sf")
library("spocc")

df <- occ(query = "solanum", from = "gbif",
          date = c("2015-01-01", "2022-12-31"),
          gbifopts = list(country = "BO"),
          has_coords = TRUE, limit = 1000)
d <- occ2df(df)
```

We use the `st_as_sf()` function to create a `sf` object called `d` that contains the `nrow(d)` = 241 locations retrieved. We set the coordinate reference system (CRS) to EPSG code 4326 since the coordinates of the locations are given by longitude and latitude values.

```
d <- st_as_sf(d[, 2:3], coords = c("longitude", "latitude"))
st_crs(d) <- "EPSG:4326"
```

In order to work with kilometers instead of degrees, we project the data to UTM 19S corresponding to the EPSG code 5356 with kilometers as units. To do that, we obtain `st_crs("EPSG:5356")$proj4string` and change `+units=m` by `+units=km`.

```
st_crs("EPSG:5356")$proj4string
projUTM <- "+proj=utm +zone=19 +south +ellps=GRS80
+towgs84=0,0,0,0,0,0,0 +units=km +no_defs"
d <- st_transform(d, crs = projUTM)
```

We also obtain the map of Bolivia with the **rnaturalearth** package, and we project it to UTM 19S with kilometers as units.

```
library(rnaturalearth)
map <- ne_countries(type = "countries", country = "Bolivia",
                    scale = "medium", returnclass = "sf")
map <- st_transform(map, crs = projUTM)
```

Figure 23.1 shows a map with the retrieved locations of *Solanum* plant species in Bolivia.

```
library("ggplot2")
ggplot() + geom_sf(data = map) +
  geom_sf(data = d) + coord_sf(datum = projUTM)
```

Finally, we create data frame `coo` with the event locations.

```
coo <- st_coordinates(d)
```

23.3.2 Prediction data

Now, we construct a matrix with the locations `coop` where we want to predict the point process intensity. To do that, we first create a raster covering the map with the `rast()` function of **terra**. Then, we retrieve the coordinates of the cells with the `xyFromCell()` function of **terra**.

```
library(sf)
library(terra)

# raster grid covering map
grid <- terra::rast(map, nrows = 100, ncols = 100)
```

```
# coordinates of all cells
xy <- terra::xyFromCell(grid, 1:ncell(grid))
```

We create a sf object called dp with the prediction locations with st_as_sf(), and use st_filter() to keep the prediction locations that lie within the map. We also retrieve the indices of the points within the map by using st_intersects() setting sparse = FALSE.

```
# transform points to a sf object
dp <- st_as_sf(as.data.frame(xy), coords = c("x", "y"),
               crs = st_crs(map))

# indices points within the map
indicespointswithin <- which(st_intersects(dp, map,
                                            sparse = FALSE))

# points within the map
dp <- st_filter(dp, map)
```

Figure 23.1 depicts the prediction locations in the study region.

```
ggplot() + geom_sf(data = map) +
  geom_sf(data = dp) + coord_sf(datum = projUTM)
```

We create the matrix coop with the prediction locations.

```
coop <- st_coordinates(dp)
```

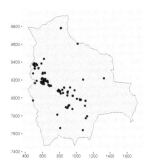 

FIGURE 23.1: Left: Locations of *Solanum* plant species in Bolivia from January 2015 to December 2022 obtained from GBIF. Right: Prediction locations in Bolivia.

23.3.3   Model

We use a LGCP to model the point pattern of plant species. Thus, we assume that the process that originates plant species locations is a Poisson process with a varying intensity expressed as

$$\log(\Lambda(\boldsymbol{s})) = \beta_0 + Z(\boldsymbol{s}),$$

where β_0 is the intercept, and $Z(\cdot)$ is a zero-mean Gaussian spatial process with Matérn covariance function.

23.3.4   Mesh construction

To fit the model using INLA and SPDE, we first construct a mesh. In the analysis of point patterns, we do not usually employ the locations as mesh vertices. We construct a mesh that covers the study region using the `inla.mesh.2d()` function setting `loc.domain` equal to a matrix with the point locations of the boundary of the region. Other arguments are as follows. `max.edge` denotes the maximum allowed triangle edge lengths in the inner region and the extension. `offset` specifies the size of the inner and outer extensions around the data locations. `cutoff` is the minimum allowed distance between points that we use to avoid building many small triangles around clustered locations. Figure 23.2 shows the mesh created.

```
library(INLA)
summary(dist(coo)) # summary of distances between event locations
```

```
   Min. 1st Qu.  Median    Mean 3rd Qu.     Max.
    0.0    66.1   197.2   240.7   384.3   1171.0
```

```
loc.d <- cbind(st_coordinates(map)[, 1], st_coordinates(map)[, 2])
mesh <- inla.mesh.2d(loc.domain = loc.d, max.edge = c(50, 100),
                  offset = c(50, 100), cutoff = 1)
```

```
plot(mesh)
points(coo, col = "red")
axis(1)
axis(2)
```

We also create variables `nv` with the number of mesh vertices, and the variable `n` with the number of events of the point pattern. Later, we will use these variables to construct the data stacks.

```
(nv <- mesh$n)
```

```
[1] 1975
```

```
(n <- nrow(coo))
```

```
[1] 241
```

We use the `inla.spde2.matern()` function to build the SPDE model on the mesh.

```
spde <- inla.spde2.matern(mesh = mesh, alpha = 2, constr = TRUE)
```

23.3.5 Observed and expected number of events

Here, we create the vectors with the observed and expected number of events. First, we create the dual mesh that consists of a set of polygons around each vertex of the original mesh (Figure 23.2). We can create the dual mesh using the `book.mesh.dual()` function that is provided in Krainski et al. (2019) and is also written at the end of this chapter.

```
dmesh <- book.mesh.dual(mesh)
plot(dmesh)
axis(1)
axis(2)
```

To fit the LGCP, the mesh vertices are considered as integration points. The expected values corresponding to the mesh vertices are proportional to the areas around the mesh vertices, that is, the areas of the polygons of the dual mesh. We calculate a vector of weights called `w` with the areas of the intersection between each polygon of the dual mesh and the study region using the following code.

```
# Domain polygon is converted into a SpatialPolygons
domain.polys <- Polygons(list(Polygon(loc.d)), '0')
domainSP <- SpatialPolygons(list(domain.polys))

# Because the mesh is larger than the study area, we need to
# compute the intersection between each polygon
# in the dual mesh and the study area
```

Constrained refined Delaunay triangulation

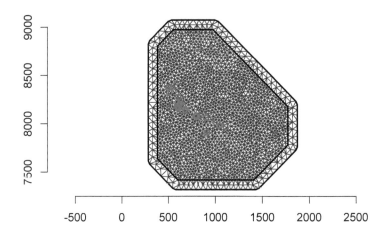

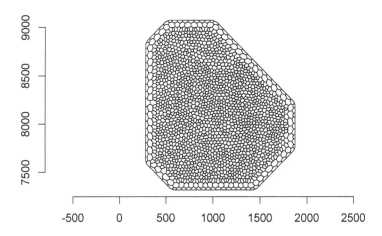

FIGURE 23.2: Mesh (top) and dual mesh (bottom) used in the SPDE approach. Event locations are depicted as red points.

```
library(rgeos)

w <- sapply(1:length(dmesh), function(i) {
  if (gIntersects(dmesh[i, ], domainSP))
    return(gArea(gIntersection(dmesh[i, ], domainSP)))
  else return(0)
})
```

Note that the sum of the weights w is equal to the area of the study region Bolivia.

```
sum(w) # sum weights
```

```
[1] 1093862
```

```
st_area(map) # area of the study region
```

```
1093862 [km^2]
```

Figure 23.3 shows the mesh together with the integration points with positive weight in black and with zero weight in red. We observe all points with zero weight are outside the study region.

```
plot(mesh)
points(mesh$loc[which(w > 0), 1:2], col = "black", pch = 20)
points(mesh$loc[which(w == 0), 1:2], col = "red", pch = 20)
```

Then, we create vectors of the augmented datasets with the observed and the expected values. The observed values will be specified in the model formula as response. The expected values will be specified in the model formula as the component E of the mean for the Poisson likelihood defined as $E_i exp(\eta_i)$, where η_i is the linear predictor.

Vector y.pp contains the response variable. The first nv elements are 0s corresponding to the mesh vertices. The last n elements are 1s corresponding to the observed events.

Vector e.pp contains the expected values. The first nv elements are the weights w representing the intersection between the areas around each of the mesh vertices and the study region. The following n elements are 0s corresponding to the point events.

Constrained refined Delaunay triangulation

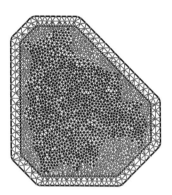

FIGURE 23.3: Mesh together with the points with positive weight in black and with zero weight in red.

```
y.pp <- rep(0:1, c(nv, n))
e.pp <- c(w, rep(0, n))
```

```
head(cbind(y.pp, e.pp))
```

```
     y.pp e.pp
[1,]   0    0
[2,]   0    0
[3,]   0    0
[4,]   0    0
[5,]   0    0
[6,]   0    0
```

```
tail(cbind(y.pp, e.pp))
```

```
        y.pp e.pp
[2211,]   1    0
[2212,]   1    0
[2213,]   1    0
```

```
[2214,]    1    0
[2215,]    1    0
[2216,]    1    0
```

23.3.6   Projection matrix

We construct the projection matrix `A.pp` to project the Gaussian random field from the observations to the triangulation vertices. This matrix is constructed using the projection matrix for the mesh vertices that is a diagonal matrix (`A.int`), and the projection matrix for the event locations (`A.y`).

```
# Projection matrix for the integration points (mesh vertices)
A.int <- Diagonal(nv, rep(1, nv))
# Projection matrix for observed points (event locations)
A.y <- inla.spde.make.A(mesh = mesh, loc = coo)
# Projection matrix for mesh vertices and event locations
A.pp <- rbind(A.int, A.y)
```

We also create the projection matrix `Ap.pp` for the prediction locations.

```
Ap.pp <- inla.spde.make.A(mesh = mesh, loc = coop)
```

23.3.7   Stack with data for estimation and prediction

Now we use the `inla.stack()` function to construct stacks for estimation and prediction that organize the data, effects, and projection matrices. In the arguments of `inla.stack()`, `data` is a list with the observed (y) and expected (e) values. Argument `A` contains the projection matrices, and argument `effects` is a list with the fixed and random effects. Then, the estimation and prediction stacks are combined in a full stack.

```
# stack for estimation
stk.e.pp <- inla.stack(tag = "est.pp",
data = list(y = y.pp, e = e.pp),
A = list(1, A.pp),
effects = list(list(b0 = rep(1, nv + n)), list(s = 1:nv)))

# stack for prediction stk.p
stk.p.pp <- inla.stack(tag = "pred.pp",
data = list(y = rep(NA, nrow(coop)), e = rep(0, nrow(coop))),
A = list(1, Ap.pp),
effects = list(data.frame(b0 = rep(1, nrow(coop))),
               list(s = 1:nv)))
```

```
# stk.full has stk.e and stk.p
stk.full.pp <- inla.stack(stk.e.pp, stk.p.pp)
```

23.3.8   Model formula and `inla()` call

The formula is specified by including the response in the left-hand side and
the random effects in the right-hand side.

```
formula <- y ~ 0 + b0 + f(s, model = spde)
```

We fit the model by calling `inla()`. In the function, we specify `link = 1` to
compute the fitted values that are given in `res$summary.fitted.values` and
`res$marginals.fitted.values` with the same link function as the `family`
specified in the model.

```
res <- inla(formula,  family = 'poisson',
data = inla.stack.data(stk.full.pp),
control.predictor = list(compute = TRUE, link = 1,
                         A = inla.stack.A(stk.full.pp)),
E = inla.stack.data(stk.full.pp)$e)
```

23.3.9   Results

A summary of the results can be inspected by typing `summary(res)`. The
data frame `res$summary.fitted.values` contains the fitted values. The indices of the rows corresponding to the predictions can be obtained with
`inla.stack.index()` specifying the tag `"pred.pp"` of the prediction stack.

```
index <- inla.stack.index(stk.full.pp, tag = "pred.pp")$data
pred_mean <- res$summary.fitted.values[index, "mean"]
pred_ll <- res$summary.fitted.values[index, "0.025quant"]
pred_ul <- res$summary.fitted.values[index, "0.975quant"]
```

Then, we add layers to the `grid` raster with the posterior mean, and 2.5 and
97.5 percentiles values in the cells that are within the map.

```
grid$mean <- NA
grid$ll <- NA
grid$ul <- NA
```

```
grid$mean[indicespointswithin] <- pred_mean
grid$ll[indicespointswithin] <- pred_ll
grid$ul[indicespointswithin] <- pred_ul
```

Finally, we create maps of the posterior mean and the lower and upper limits of 95% credible intervals of the intensity of the point process of species in Bolivia (Figure 23.4). To do that, we use the `levelplot()` function of the **rasterVis** package specifying `names.attr` with the name of each panel and `layout` with the number of columns and rows.

```
library(rasterVis)
levelplot(raster::brick(grid), layout = c(3, 1),
names.attr = c("Mean", "2.5 percentile", "97.5 percentile"))
```

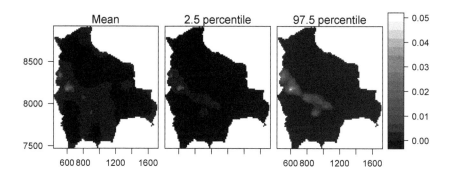

FIGURE 23.4: Maps with the posterior mean of the intensity of the point process of species in Bolivia (left), and lower (center) and upper (right) limits of 95% credible intervals.

Overall, we observe a low intensity of species, with higher intensity in the central west part of Bolivia. Note that we have modeled species occurrence data retrieved from GBIF by assuming the observed spatial point pattern is a realization of the underlying process that generates the species locations. In real applications, it is important to understand how data was collected, and assess potential data biases such as overrepresentation of certain areas that can invalidate the results. Moreover, it is important to incorporate expert knowledge, to create models that include relevant covariates and random effects to account for various types of variability, enabling a more comprehensive understanding of the variable under investigation.

23.3.10   Function to create the dual mesh

The following code corresponds to the `book.mesh.dual()` function to create
the dual mesh.

```
book.mesh.dual <- function(mesh) {
    if (mesh$manifold=='R2') {
        ce <- t(sapply(1:nrow(mesh$graph$tv), function(i)
            colMeans(mesh$loc[mesh$graph$tv[i, ], 1:2])))
        library(parallel)
        pls <- mclapply(1:mesh$n, function(i) {
            p <- unique(Reduce('rbind', lapply(1:3, function(k) {
            j <- which(mesh$graph$tv[,k]==i)
            if (length(j)>0)
            return(rbind(ce[j, , drop=FALSE],
            cbind(mesh$loc[mesh$graph$tv[j, k], 1] +
            mesh$loc[mesh$graph$tv[j, c(2:3,1)[k]], 1],
            mesh$loc[mesh$graph$tv[j, k], 2] +
            mesh$loc[mesh$graph$tv[j, c(2:3,1)[k]], 2])/2))
            else return(ce[j, , drop=FALSE])
            })))
            j1 <- which(mesh$segm$bnd$idx[,1]==i)
            j2 <- which(mesh$segm$bnd$idx[,2]==i)
            if ((length(j1)>0) | (length(j2)>0)) {
            p <- unique(rbind(mesh$loc[i, 1:2], p,
            mesh$loc[mesh$segm$bnd$idx[j1, 1], 1:2]/2 +
            mesh$loc[mesh$segm$bnd$idx[j1, 2], 1:2]/2,
            mesh$loc[mesh$segm$bnd$idx[j2, 1], 1:2]/2 +
            mesh$loc[mesh$segm$bnd$idx[j2, 2], 1:2]/2))
            yy <- p[,2]-mean(p[,2])/2-mesh$loc[i, 2]/2
            xx <- p[,1]-mean(p[,1])/2-mesh$loc[i, 1]/2
            }
            else {
            yy <- p[,2]-mesh$loc[i, 2]
            xx <- p[,1]-mesh$loc[i, 1]
            }
            Polygon(p[order(atan2(yy,xx)), ])
        })
        return(SpatialPolygons(lapply(1:mesh$n, function(i)
            Polygons(list(pls[[i]]), i))))
    }
    else stop("It only works for R2!")
}
```

A

The R software

A.1 R and RStudio

R[1] is a free and open-source software environment for statistical computing and graphics that provides many excellent packages for importing and manipulating data, statistical analysis, and visualization. R can be downloaded and installed from CRAN (Comprehensive R Archive Network)[2]. R can be run using the integrated development environment (IDE) called RStudio which can be freely downloaded from https://posit.co/download/rstudio-desktop/. RStudio includes a console, a syntax-highlighting editor for writing and editing R code, and a variety of tools for data visualization, debugging, and management of files and R projects.

The RStudio IDE has typically four panes (Figure A.1). In the code editor pane (top-left), we create and view the R script files. In the console pane (bottom-left), we see the execution and the output of the R code. To interact with R, we can type commands in the console or write code in script files in the code editor and copy-paste commands to the console. The Environment/History pane (top-right) contains tabs with datasets, variables, and other R objects, as well as the history of the previous R commands executed. This pane may also contain Git options for version control. Finally, the Files/Plots/Packages/Help pane (bottom-right) allows us to see the files in our working directory, the graphs generated, as well as packages and help pages.

A.2 Installation of R packages

R provides functionality to read and write data; create R objects such as vectors, matrices, data frames and lists; conduct statistical analyses and plotting. We can also install additional R packages for data retrieval, manipulation, analysis, visualization, and reporting. To install an R package from CRAN, we use the

---

[1]https://www.r-project.org
[2]https://cran.rstudio.com

A The R software

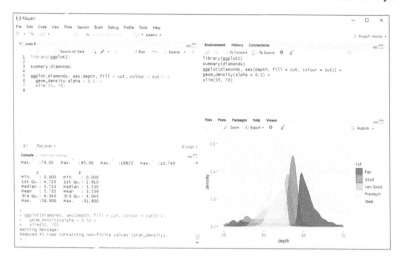

FIGURE A.1: Screenshot of RStudio.

`install.packages()` function passing the name of the package. Then, to use the package, we load it with `library()`. For example, we can install and load the **sf** package (Pebesma, 2022a) to work with spatial vector data as follows:

```
install.packages("sf")
library(sf)
```

A.3   Packages for data visualization

A.3.1   ggplot2

The **ggplot2** package (Wickham et al., 2022a) uses a grammar of graphics which defines the rules of structuring mathematic and aesthetic elements to build graphs layer-by-layer. To create a plot with **ggplot2**, we call `ggplot()` specifying the data frame with the variables to plot (`data`), and the aesthetic mappings between variables in the data and visual properties of the objects in the graph, such as the position and color of points or lines (`mapping = aes()`). Then, we use the + symbol to add layers of graphical components to the graph. Layers consist of geoms, stats, scales, coords, facets, and themes. For example, we add objects to the graph with `geom_*()` functions (e.g., `geom_point()` for points, `geom_line()` for lines). We can also add color scales (e.g., `scale_colour_brewer()`), faceting specifications (e.g., `facet_wrap()`), and coordinate systems (e.g., `coord_flip()`). To save a plot, we use `ggsave()`.

Here, we use the `st_read()` function of **sf** to read the shapefile that contains the number of sudden infant deaths in North Carolina, USA, in 1974, and create a map using **ggplot2** (Figure A.2).

```
library(ggplot2)
library(sf)
library(viridis)

d <- st_read(system.file("shape/nc.shp", package = "sf"),
             quiet = TRUE)
ggplot(d) + geom_sf(aes(fill = SID74)) +
  scale_fill_viridis() + theme_bw()
```

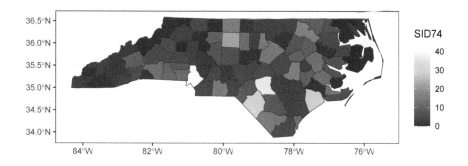

FIGURE A.2: Map of number of sudden infant deaths in North Carolina, USA, in 1974 created with **ggplot2**.

A.3.2 HTML widgets

HTML widgets[3] are interactive web visualizations built with JavaScript. Here, we provide examples of HTML widgets that allow us to create interactive maps, time series plots and tables. Other examples of HTML widgets can be seen at this website[4]. The **leaflet** package (Cheng et al., 2022a) allows us to create interactive maps supporting panning and zooming. We can include basemaps using map tiles to put data into context. A set of available basemaps can be seen here[5]. Figure A.3 shows a map created with **leaflet** of the locations and Richter magnitude of seismic events occurred near Fiji in 1964 that are contained in the `quakes` data.

---

[3]http://www.htmlwidgets.org/
[4]https://www.htmlwidgets.org/showcase_leaflet.html
[5]http://leaflet-extras.github.io/leaflet-providers/preview/index.html

```
library(leaflet)
library(sf)

d <- quakes[1:20, ]
pal <- colorNumeric(palette = "YlOrRd", domain = d$mag)
leaflet(d) %>% addTiles() %>%
  addCircleMarkers(color = ~ pal(mag)) %>%
  leaflet::addLegend(pal = pal, values = ~ mag)
```

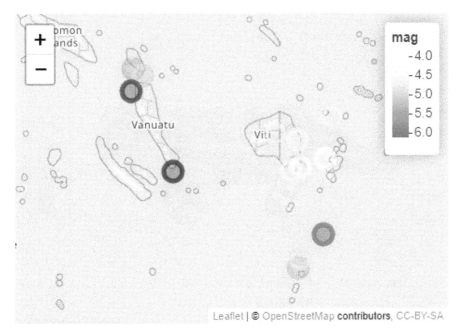

FIGURE A.3: Map of the locations and Richter magnitude of seismic events occurred near Fiji in 1964 created with **leaflet**.

The **dygraphs** package (Vanderkam et al., 2018) provides functionality to create interactive plots of time series data. Figure A.4 shows a time series plot created with **dygraphs** with the mean annual temperature in degrees Fahrenheit in New Haven, Connecticut, USA, over the years contained in the nhtemp data.

```
library(dygraphs)
dygraph(nhtemp, main = "New Haven Temperatures") %>%
  dyRangeSelector(dateWindow = c("1920-01-01", "1960-01-01"))
```

The **DT** package (Xie et al., 2022a) allows us to display matrices and data

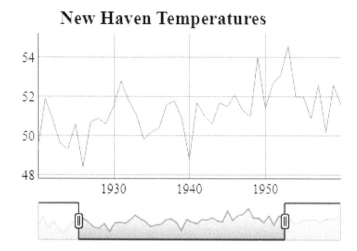

FIGURE A.4: Time series plot of the mean annual temperature in degrees Fahrenheit in New Haven, Connecticut, USA, created with **dygraphs**.

frames as tables supporting filtering, pagination and sorting. For example, Figure A.5 depicts a table created with **DT** showing the names, and the sepal and petal lengths and widths in centimeters of 150 flowers contained in the `iris` dataset.

```
library(DT)
datatable(iris, options = list(pageLength = 5))
```

A.4 Reproducible reports and dashboards

A.4.1 R Markdown

The package **rmarkdown** (Allaire et al., 2022) allows us to easily turn our analyses into fully reproducible documents that can be shared with others in a variety of formats including HTML and PDF. An R Markdown document is a text file with extension .Rmd that intermingles text and R code, and can include narrative text, tables, and visualizations. When the R Markdown document is compiled, the R code is executed and a report with the output of the R code is created. Resources to learn R Markdown include Xie et al. (2022b), Xie et al. (2018), and chapter 11 of Moraga (2019) which provides a

	Sepal.Length	Sepal.Width	Petal.Length	Petal.Width	Species
1	5.1	3.5	1.4	0.2	setosa
2	4.9	3	1.4	0.2	setosa
3	4.7	3.2	1.3	0.2	setosa
4	4.6	3.1	1.5	0.2	setosa
5	5	3.6	1.4	0.2	setosa

Show 5 entries   Search:

Showing 1 to 5 of 150 entries

Previous   1   2   3   4   5   ...   30   Next

FIGURE A.5: Table created with **DT** showing the information of the `iris` dataset.

reproducible example of how to create an R Markdown document that includes an exploratory data analysis with tables and visualizations.

A new R Markdown document (`.Rmd`) can be created by clicking `File > New File > R Markdown` in RStudio. From the `.Rmd` file, a report can be generated using the `Knit` button in RStudio or executing `rmarkdown::render("name.Rmd", "output_document")`, where `name.Rmd` is the name of the `.Rmd` file, and `"output_document"` the type of output (e.g., `"html_document"`, `"pdf_document"`). Note that LaTeX is needed to generate PDF documents. The LaTeX distribution TinyTeX can be installed with the **tinytex** package (Xie, 2022) with `tinytex::install_tinytex()` (Xie et al., 2022b). Alternatively, LaTeX can be installed using the resources in the https://www.latex-project.org/get/ website.

An R Markdown document includes three basic components, namely, a YAML header, Markdown text, and R code chunks. At the beginning of the document, we write a YAML header surrounded by `---` that indicates several options such as title, author, date, and type of output file.

```
---
title: "Report"
author: "Paula Moraga"
date: 1 July 2023
```

```
output: html_document
---
```

The text is written in Markdown syntax. For example, we can use asterisks for italic text (`*text*`) and double asterisks for bold text (`**text**`) . We can also include equations in LaTeX.

The R code is written within R code chunks which start with three backticks `` ```{r} `` and end with three backticks `` ``` ``. R code chunks can be specified using several options like `echo=FALSE` to hide code and `warning=FALSE` to supress warnings.

```
```{r, warning = FALSE}
# R code to be executed
```
```

We can include images using `knitr::include_graphics("path/img.png")` and tables created with `knitr::kable()`. We can also include HTML widgets[6] such as objects created with **leaflet**, **DT**, and **dygraphs**.

A.4.2   Quarto

Quarto (Allaire, 2022) is a multi-language, next-generation version of R Markdown, that includes many new features and capabilities. A Quarto document has extension `.qmd` and can be rendered as formats like PDF and Word using the **Render** button of RStudio or typing `quarto::quarto_render("name.qmd")` in the console. Quarto is able to render most existing `.Rmd` files without modification. Quarto documents are formed of a YAML header, Markdown text, and R code chunks. The R code chunks options are identified by `#|` at the beginning of the lines. For example,

```
```{r, warning=FALSE}
#| label: load-packages
#| include: false

plot(1:10, 1:10)
```
```

A.4.3   Flexdashboard

The **flexdashboard** package (Sievert et al., 2022a) allows us to create dashboards in HTML format that contain several related data visualizations. Ex-

---

[6]https://www.htmlwidgets.org/

amples of dashboards created with **flexdashboard** can be seen at the RStudio website[7]. Chapter 12 of Moraga (2019) explains how to build a flexdashboard with several components showing air pollution levels in each of the world's countries (Figure A.6).

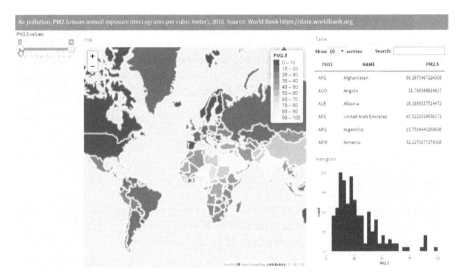

FIGURE A.6: Screenshot of a flexdashboard to visualize air pollution data globally.

To create a flexdashboard, we need to write an R Markdown file with extension `.Rmd`. The YAML header of the flexdashboard document needs to have the option `output: flexdashboard::flex_dashboard`. Dashboard components are shown according to a layout that specifies the columns and rows. Columns are included with --------------, and rows for each column with `###`. Layouts can also be specified row-wise rather than column-wise by adding `orientation: rows` in the YAML. Layout examples including tabs, multiple pages, and sidebars can be seen at the R Markdown website[8].

The R code to create the dashboard's visualizations is written within R code chunks. Dashboards can contain a wide variety of components including images, tables, equations, and HTML widgets. They can also contain value boxes[9] to display single values with titles and icons, and gauges[10] that display values on a meter within a specified range. Moreover, it is also possible to include navigation bars with links to social media, source code, or other links related to the dashboard.

---

[7]https://pkgs.rstudio.com/flexdashboard/articles/examples.html
[8]https://pkgs.rstudio.com/flexdashboard/articles/layouts.html
[9]https://bookdown.org/yihui/rmarkdown/dashboard-components.html#value-boxes
[10]https://bookdown.org/yihui/rmarkdown/dashboard-components.html#gauges

A.4.4 Shiny

Shiny (Chang et al., 2022) is a web application framework for R that enables to build interactive web applications. Examples of Shiny apps can be seen at https://shiny.posit.co/r/gallery/. The **SpatialEpiApp** package (Moraga, 2018b) contains a Shiny app for disease risk estimation, cluster detection, and interactive visualization. Chapters 13-15 of Moraga (2019) provide an introduction to Shiny as well as examples to build a Shiny app to upload and visualize spatial and spatio-temporal data.

A Shiny app can be built by creating a directory that contains an R file with three components. Namely, a `ui` user interface object which controls layout and appearance of the app, a `server()` function with instructions to build objects displayed in the ui, and a call to `shinyApp()` that creates the Shiny app from the `ui`/`server()` pair.

```
# define user interface object
ui <- fluidPage( )
# define server() function
server <- function(input, output){ }
# call to shinyApp() which returns the Shiny app
shinyApp(ui = ui, server = server)
```

In the `ui` object, we can include input objects that allow us to interact with the app by modifying their values (e.g., texts, dates, files), and output objects we want to show in the app (e.g., texts, tables, plots). The `server()` function contains the R code to build the outputs. If this code uses an input value, the output will be rebuilt whenever the value changes creating reactivity. The app directory can also contain data or other R scripts needed by the app. We can also write two separate files `ui.R` and `server.R` for an easier management of code in large apps.

There are two options to share a Shiny app. We can share the R scripts with other users so they can launch the app from R with the `runApp()` function specifying the path of the directory of the app. Another sharing option that does not require the users to have R is to host the app as a web page at its own URL so the app can be navigated through the internet with a web browser.

Bibliography

Allaire, J. (2022). *quarto: R Interface to 'Quarto' Markdown Publishing System*. R package version 1.2.

Allaire, J., Xie, Y., McPherson, J., Luraschi, J., Ushey, K., Atkins, A., Wickham, H., Cheng, J., Chang, W., and Iannone, R. (2022). *rmarkdown: Dynamic Documents for R*. R package version 2.19.

Anselin, L. (1995). Local Indicators of Spatial Association - LISA. *Geographical Analysis*, 27:93–115.

Appelhans, T. and Detsch, F. (2021). *leafpop: Include Tables, Images and Graphs in Leaflet Pop-Ups*. R package version 0.1.0.

Appelhans, T., Detsch, F., Reudenbach, C., and Woellauer, S. (2022). *mapview: Interactive Viewing of Spatial Data in R*. R package version 2.11.0.

Appelhans, T. and Russell, K. (2019). *leafsync: Small Multiples for Leaflet Web Maps*. R package version 0.1.0.

Baddeley, A., Rubak, E., and Turner, R. (2016). *Spatial Point Patterns. Methodology and Applications with R*. Chapman & Hall/CRC.

Baddeley, A., Turner, R., and Rubak, E. (2022). *spatstat: Spatial Point Pattern Analysis, Model-Fitting, Simulation, Tests*. R package version 3.0-2.

Banerjee, S., Carlin, B. P., and Gelfand, A. E. (2004). *Hierarchical Modeling and Analysis for Spatial Data*. Chapman & Hall/CRC. ISBN 978-1439819173.

Besag, J., York, J., and Mollié, A. (1991). Bayesian image restoration with applications in spatial statistics (with discussion). *Annals of the Institute of Statistical Mathematics*, 43:1–59.

Bivand, R. (2022). *spdep: Spatial Dependence: Weighting Schemes, Statistics*. R package version 1.2-7.

Bivand, R., Keitt, T., and Rowlingson, B. (2023). *rgdal: Bindings for the Geospatial Data Abstraction Library*. R package version 1.6-4.

Bivand, R. and Lewin-Koh, N. (2022). *maptools: Tools for Handling Spatial Objects*. R package version 1.1-6.

Bivand, R., Nowosad, J., and Lovelace, R. (2022). *spData: Datasets for Spatial Analysis*. R package version 2.2.1.

Bivand, R. and Rundel, C. (2022). *rgeos: Interface to Geometry Engine - Open Source (GEOS)*. R package version 0.6-1.

Boyandin, I. (2023). *flowmapblue: Flow map rendering*. R package version 0.0.1.

Brownrigg, R. (2022). *maps: Draw Geographical Maps*. R package version 3.4.1.

Cameletti, M., Lindgren, F., Simpson, D., and Rue, H. (2013). Spatio-temporal modeling of particulate matter concentration through the spde approach. *AStA Advances in Statistical Analysis*, 97(2):109–131.

Carslaw, D., Davison, J., and Ropkins, K. (2023). *openair: Tools for the Analysis of Air Pollution Data*. R package version 2.17-0.

Chamberlain, S. (2021). *spocc: Interface to Species Occurrence Data Sources*. R package version 1.2.0.

Chang, W. (2022a). *webshot: Take Screenshots of Web Pages*. R package version 0.5.4.

Chang, W. (2022b). *webshot2: Take Screenshots of Web Pages*. R package version 0.1.0.

Chang, W., Cheng, J., Allaire, J., Sievert, C., Schloerke, B., Xie, Y., Allen, J., McPherson, J., Dipert, A., and Borges, B. (2022). *shiny: Web Application Framework for R*. R package version 1.7.4.

Chang, W. and Schloerke, B. (2022). *chromote: Headless Chrome Web Browser Interface*. R package version 0.1.1.

Cheng, J., Karambelkar, B., and Xie, Y. (2022a). *leaflet: Create Interactive Web Maps with the JavaScript Leaflet Library*. R package version 2.1.1.

Cheng, J., Sievert, C., Schloerke, B., Chang, W., Xie, Y., and Allen, J. (2022b). *htmltools: Tools for HTML*. R package version 0.5.4.

Cressie, N. A. C. (1993). *Statistics for Spatial Data*. John Wiley & Sons, New York. ISBN 978-0471002550.

Davies, T. M. and Marshall, J. C. (2023). *sparr: Spatial and Spatiotemporal Relative Risk*. R package version 2.3-10.

de Sousa, K., Sparks, A. H., and Ghosh, A. (2022). *chirps: API Client for CHIRPS and CHIRTS*. R package version 0.1.4.

Diggle, P. J. (2014). *Statistical analysis of spatial and spatio-temporal point patterns*. CRC Press.

Diggle, P. J., Menezes, R., and li Su, T. (2010). Geostatistical inference under preferential sampling. *Journal of the Royal Statistical Society: Series C*, 59:191–232.

Diggle, P. J., Moraga, P., Rowlingson, B., and Taylor, B. M. (2013). Spatial and Spatio-Temporal Log-Gaussian Cox Processes: Extending the Geostatistical Paradigm. *Statistical Science*, 28(4):542–563.

Diggle, P. J. and Ribeiro Jr., P. J. (2007). *Model-based Geostatistics*. Springer Series in Statistics, 1st edition. ISBN 978-0387329079.

Diggle, P. J., Tawn, J. A., and Moyeed, R. A. (1998). Model-based geostatistics. *Journal of the Royal Statistical Society. Series C (Applied Statistics)*, 47(3):299–350.

Freni-Sterrantino, A., Ventrucci, M., and Rue, H. (2018). A note on intrinsic conditional autoregressive models for disconnected graphs. *Spatial and Spatio-temporal Epidemiology*, (26):25–34.

Garnier, S. (2021). *viridis: Colorblind-Friendly Color Maps for R*. R package version 0.6.2.

Gatrell, A. C., Bailey, T. C., Diggle, P. J., and Rowlingson, B. S. (1996). Spatial Point Pattern Analysis and Its Application in Geographical Epidemiology. *Transactions of the Institute of British Geographers*, 21(1):256–274.

Geary, R. C. (1954). The Contiguity Ratio and Statistical Mapping. *The Incorporated Statistician*, 5(3):115–146.

Gelman, A., Carlin, J. B., Stern, H. S., Dunson, D. B., Vehtari, A., and Rubin, D. B. (2013). *Bayesian Data Analysis*. Chapman and Hall/CRC.

Gómez, M. J., Barboza, L. A., Vásquez, P., and Moraga, P. (2023). Bayesian spatial modeling of childhood overweight and obesity prevalence in Costa Rica. *BMC Public Health*, 23(1):1–11.

Gómez-Rubio, V. (2020). *Bayesian inference with INLA*. Chapman & Hall/CRC, Boca Raton, Florida.

González, J. A. and Moraga, P. (2022). An adaptive kernel estimator for the intensity function of spatio-temporal point processes. arXiv:2208.12026 [stat.ME].

González, J. A. and Moraga, P. (2023a). A multitype Fiksel interaction model for tumour immune microenvironments. arXiv:2307.05556 [stat.AP].

González, J. A. and Moraga, P. (2023b). Non-parametric analysis of spatial and spatio-temporal point patterns. *The R Journal*, 15:65–82.

González, J. A. and Moraga, P. (2023c). On adaptive kernel intensity estimation on linear networks. arXiv:2309.09303 [stat.ME].

Gotway, C. A. and Young, L. J. (2002). Combining incompatible spatial data. *Journal of the American Statistical Association*, 97(458):632–648.

Grolemund, G. (2014). *Hands-On Programming with R.* O'Reilly, Sebastopol, California, 1st edition. ISBN 978-1449359010.

Hagan, J. E., Moraga, P., Costa, F., Capian, N., Ribeiro, G. S., Jr., E. A. Wunder, Felzemburgh, R. D. M., Reis, R. B., Nery, N., Santana, F. S., Fraga, D., dos Santos, B. L., Santos, A. C., Queiroz, A., Tassinari, W., Carvalho, M. S., Reis, M. G., Diggle, P. J., and Ko, A. I. (2016). Spatio-temporal determinants of urban leptospirosis transmission: Four-year prospective cohort study of slum residents in Brazil. *Public Library of Science: Neglected Tropical Diseases*, 10(1):e0004275.

Held, L., Schrödle, B., and Rue, H. (2010). Posterior and cross-validatory predictive checks: A comparison of mcmc and inla. In Kneib, T. and Tutz, G., editors, *Statistical Modelling and Regression Structures – Festschrift in Honour of Ludwig Fahrmeir*, pages 91–110. Springer Verlag, Berlin.

Hernangomez, D. (2022). *mapSpain: Administrative Boundaries of Spain.* R package version 0.7.0.

Hernangomez, D. (2023a). *giscoR: Download Map Data from GISCO API - Eurostat.* R package version 0.3.4.

Hernangomez, D. (2023b). *tidyterra: tidyverse Methods and ggplot2 Helpers for terra Objects.* R package version 0.4.0.

Hijmans, R. J. (2022). *terra: Spatial Data Analysis.* R package version 1.6-47.

Hijmans, R. J. (2023). *raster: Geographic Data Analysis and Modeling.* R package version 3.6-13.

Hijmans, R. J., Barbosa, M., Ghosh, A., and Mandel, A. (2023). *geodata: Download Geographic Data.* R package version 0.5-7.

Hijmans, R. J., Phillips, S., Leathwick, J., and Elith, J. (2022). *dismo: Species Distribution Modeling.* R package version 1.3-9.

Hollister, J. (2022). *elevatr: Access Elevation Data from Various APIs.* R package version 0.4.2.

Illian, J. B., Sørbye, S. H., Rue, H., and Hendrichsen, D. (2012). Using INLA To Fit A Complex Point Process Model With Temporally Varying Effects - A Case Study. *Journal of Environmental Statistics*, 3.

Kahle, D., Wickham, H., and Jackson, S. (2022). *ggmap: Spatial Visualization with ggplot2.* R package version 3.0.1.

Kim, A. Y., Wakefield, J., and Moise, M. (2021). *SpatialEpi: Methods and Data for Spatial Epidemiology.* R package version 1.2.7.

Krainski, E. T., Gómez-Rubio, V., Bakka, H., Lenzi, A., Castro-Camilo, D., Simpson, D., Lindgren, F., and Rue, H. (2019). *Advanced Spatial Modeling with Stochastic Partial Differential Equations Using R and INLA*. Chapman & Hall/CRC, Boca Raton, Florida, 1st edition. ISBN 978-1138369856.

Lawson, A. B., Rotejanaprasert, C., Moraga, P., and Choi, J. (2015). A shared neighbor conditional autoregressive model for small area spatial data. *Environmetrics*, 26(6):383–392.

Leasure, D. R., Bondarenko, M., Darin, E., and Tatem, A. J. (2023). *wopr: An R Package to access the WorldPop Open Population Repository (WOPR)*. R package version 1.3.3.

Lindgren, F. and Rue, H. (2015). Bayesian Spatial Modelling with R-INLA. *Journal of Statistical Software*, 63.

Mahmood, M., Ribeiro Amaral, A. V., Mateu, J., and Moraga, P. (2022). Modeling infectious disease dynamics: Integrating contact tracing-based stochastic compartment and spatio-temporal risk models. *Spatial Statistics*, 51:100691.

Matheron, G. (1963). Principles of geostatistics. *Economic Geology*, 58:1246–1266.

Matheson, J. E. and Winkler, R. L. (1976). Scoring rules for continuous probability distributions. *Management Science*, 22(10):1087–1096.

Moraga, P. (2017). SpatialEpiApp: A Shiny Web Application for the analysis of Spatial and Spatio-Temporal Disease Data. *Spatial and Spatio-temporal Epidemiology*, 23:47–57.

Moraga, P. (2018a). Small Area Disease Risk Estimation and Visualization Using R. *The R Journal*, 10(1):495–506.

Moraga, P. (2018b). *SpatialEpiApp: A Shiny Web Application for the Analysis of Spatial and Spatio-Temporal Disease Data*. R package version 0.5.

Moraga, P. (2019). *Geospatial Health Data: Modeling and Visualization with R-INLA and Shiny*. Chapman & Hall/CRC Biostatistics Series, Boca Raton, Florida.

Moraga, P. (2021a). Handbook of Spatial Epidemiology. *Journal of the American Statistical Association*, 116(533):451–453.

Moraga, P. (2021b). Species Distribution Modeling using Spatial Point Processes: a Case Study of Sloth Occurrence in Costa Rica. *The R Journal*, 12(2):293–310.

Moraga, P. and Baker, L. (2022). rspatialdata: a collection of data sources and tutorials on downloading and visualising spatial data using R. *F1000Research*, 11:770.

Moraga, P., Cramb, S., Mengersen, K., and Pagano, M. (2017). A geostatistical model for combined analysis of point-level and area-level data using INLA and SPDE. *Spatial Statistics*, 21:27–41.

Moraga, P., Dean, C., Inoue, J., Morawiecki, P., Noureen, S. R., and Wang, F. (2021). Bayesian spatial modelling of geostatistical data using INLA and SPDE methods: A case study predicting malaria risk in Mozambique. *Spatial and Spatio-temporal Epidemiology*, 39:100440.

Moraga, P., Dorigatti, I., Kamvar, Z. N., Piatkowski, P., Toikkanen, S. E., VP, Naraj, Donnelly, C. A., and Jombart, T. (2019). epiflows: an R package for risk assessment of travel-related spread of disease. *F1000Research*, 7:1374.

Moraga, P. and Kulldorff (2016). Detection of spatial variations in temporal trends with a quadratic function. *Statistical Methods for Medical Research*, 25(4):1422–1437.

Moraga, P. and Lawson, A. B. (2012). Gaussian component mixtures and CAR models in Bayesian disease mapping. *Computational Statistics & Data Analysis*, 56(6):1417–1433.

Moraga, P. and Montes, F. (2011). Detection of spatial disease clusters with LISA functions. *Statistics in Medicine*, 30:1057–1071.

Moraga, P. and Ozonoff, A. (2013). Model-based imputation of missing data from the 122 Cities Mortality Reporting System (122 CMRS). *Stochastic Environmental Research and Risk Assessment*, 29(5):1499–1507.

Moran, P. A. P. (1950). Notes on Continuous Stochastic Phenomena. *Biometrika*, 37(1/2):17–23.

Neuwirth, E. (2022). *RColorBrewer: ColorBrewer Palettes*. R package version 1.1-3.

Openshaw, S. (1984). *The Modifiable Areal Unit Problem*. Geo Books, Norwich, UK. ISBN 978-0860941347.

Padgham, M., Rudis, B., Lovelace, R., Salmon, M., and Maspons, J. (2023). *osmdata: Import OpenStreetMap Data as Simple Features or Spatial Objects*. R package version 0.2.0.

Pavani, J., Bastos, L. S., and Moraga, P. (2023). Joint spatial modeling of the risks of co-circulating mosquito-borne diseases in Ceará, Brazil. *Spatial and Spatio-temporal Epidemiology*, 47:100616.

Pebesma, E. (2022a). *sf: Simple Features for R*. R package version 1.0-9.

Pebesma, E. (2022b). *stars: Spatiotemporal Arrays, Raster and Vector Data Cubes*. R package version 0.6-0.

Pebesma, E. and Bivand, R. (2022). *sp: Classes and Methods for Spatial Data*. R package version 1.5-1.

Pebesma, E. and Graeler, B. (2022). *gstat: Spatial and Spatio-Temporal Geostatistical Modelling, Prediction and Simulation.* R package version 2.1-0.

Pedersen, T. L. (2022). *patchwork: The Composer of Plots.* R package version 1.1.2.

Pedersen, T. L. and Robinson, D. (2022). *gganimate: A Grammar of Animated Graphics.* R package version 1.0.8.

Pereira, R. H. M. and Goncalves, C. N. (2022). *geobr: Download Official Spatial Data Sets of Brazil.* R package version 1.7.0.

Pfeffer, D., Lucas, T., May, D., Keddie, S., Rozier, J., Watson, O., and Gibson, H. (2020). *malariaAtlas: An R Interface to Open-Access Malaria Data, Hosted by the Malaria Atlas Project.* R package version 1.0.1.

Piatkowski, P., Moraga, P., Jombart, T., Nagraj, V., Kamvar, Z. N., and Toikkanen, S. E. (2018). *epiflows: Predicting Disease Spread from Flow Data.* R package version 0.2.0.

Piburn, J. (2020). *wbstats: Programmatic Access to Data and Statistics from the World Bank API.* R package version 1.0.4.

Pierce, D. (2023). *ncdf4: Interface to Unidata netCDF (Version 4 or Earlier) Format Data Files.* R package version 1.21.

Possenriede, D., Sadler, J., and Salmon, M. (2021). *opencage: Geocode with the OpenCage API.* R package version 0.2.2.

Ribeiro Amaral, A. V., González, J. A., and Moraga, P. (2023a). Spatio-temporal modeling of infectious diseases by integrating compartment and point process models. *Stochastic Environmental Research and Risk Assessment*, 37:1519–1533.

Ribeiro Amaral, A. V., Krainski, E. T., Zhong, R., and Moraga, P. (2023b). Model-Based Geostatistics Under Spatially Varying Preferential Sampling. *Journal of Agricultural, Biological, and Environmental Statistics.* doi: 10.1007/s13253-023-00571-0.

Ribeiro Jr, P. J., Diggle, P., Christensen, O., Schlather, M., Bivand, R., and Ripley, B. (2022). *geoR: Analysis of Geostatistical Data.* R package version 1.9-2.

Riebler, A., Sørbye, S. H., Simpson, D., and Rue, H. (2016). An intuitive Bayesian spatial model for disease mapping that accounts for scaling. *Statistical Methods in Medical Research*, 25(4):1145–1165.

Robinson, W. S. (1950). Ecological Correlations and the Behavior of Individuals. *American Sociological Review*, 15(3):351–357.

Rue, H., Lindgren, F., and Teixeira Krainski, E. (2022). *INLA: Full Bayesian*

Analysis of Latent Gaussian Models using Integrated Nested Laplace Approximations. R package version 22.12.16.

Rue, H., Martino, S., and Chopin, N. (2009). Approximate Bayesian inference for latent Gaussian models using integrated nested Laplace approximations (with discussion). *Journal of the Royal Statistical Society B*, 71:319–392.

Sassi, G. (2021). *geoFourierFDA: Ordinary Functional Kriging Using Fourier Smoothing and Gaussian Quadrature*. R package version 0.1.0.

Schlather, M., Malinowski, A., Menck, P. J., Oesting, M., and Strokorb, K. (2015). Analysis, simulation and prediction of multivariate random fields with package randomfields. *Journal of Statistical Software*, 63(8):1–15.

Schloerke, B., Cook, D., Larmarange, J., Briatte, F., Marbach, M., Thoen, E., Elberg, A., and Crowley, J. (2021). *GGally: Extension to ggplot2*. R package version 2.1.2.

Sebastian, G. (2023). *leaflet.extras2: Extra Functionality for leaflet Package*. R package version 1.2.1.

Sievert, C., Iannone, R., Allaire, J., and Borges, B. (2022a). *flexdashboard: R Markdown Format for Flexible Dashboards*. R package version 0.6.0.

Sievert, C., Parmer, C., Hocking, T., Chamberlain, S., Ram, K., Corvellec, M., and Despouy, P. (2022b). *plotly: Create Interactive Web Graphics via plotly.js*. R package version 4.10.1.

Silverman, B. W. (1986). *Density Estimation for Statistics and Data Analysis*. Chapman & Hall/CRC.

Simpson, D., Illian, J. B., Lindgren, F., Sørbye, S. H., and Rue, H. (2016). Going off grid: computationally efficient inference for log-Gaussian Cox processes. *Biometrika*, 103(1):49–70.

Simpson, D., Rue, H., Riebler, A., Martins, T. G., and Sørbye, S. H. (2017). Penalising model component complexity: A principled, practical approach to constructing priors. *Statistical Science*, 32:1–28.

South, A. (2017). *rnaturalearth: World Map Data from Natural Earth*. R package version 0.1.0.

South, A. (2023). *rnaturalearthhires: High Resolution World Vector Map Data from Natural Earth used in rnaturalearth*. https://docs.ropensci.org/rnaturalearthhires.

Spiegelhalter, D. J., Best, N. G., Carlin, B. P., and van der Linde, A. (2002). Bayesian measures of model complexity and fit (with discussion). *Journal of the Royal Statistical Society, Series B*, 64:583–616.

Tennekes, M. (2022). *tmap: Thematic Maps*. R package version 3.3-3.

Tobler, W. R. (1970). A Computer Movie Simulating Urban Growth in the Detroit Region. *Economic Geography*, 46:234–240.

Vaidyanathan, R., Xie, Y., Allaire, J., Cheng, J., Sievert, C., and Russell, K. (2023). *htmlwidgets: HTML Widgets for R*. R package version 1.6.1.

Valavi, R., Elith, J., Lahoz-Monfort, J., Flint, I., and Guillera-Arroita, G. (2023). *blockCV: Spatial and Environmental Blocking for K-Fold and LOO Cross-Validation*. R package version 3.1-2.

Vanderkam, D., Allaire, J., Owen, J., Gromer, D., and Thieurmel, B. (2018). *dygraphs: Interface to Dygraphs Interactive Time Series Charting Library*. R package version 1.1.1.6.

Wadoux, A. M.-C., Heuvelink, G. B., de Bruin, S., and Brus, D. J. (2021). Spatial cross-validation is not the right way to evaluate map accuracy. *Ecological Modelling*, 457:109692.

Walker, K. (2023). *tigris: Load Census TIGER/Line Shapefiles*. R package version 2.0.1.

Walker, K. and Herman, M. (2023). *tidycensus: Load US Census Boundary and Attribute Data as tidyverse and sf-Ready Data Frames*. R package version 1.3.2.

Wang, X., Ryan, Y. Y., and Faraway, J. J. (2018). *Bayesian Regression Modeling with INLA*. Chapman & Hall/CRC, Boca Raton, Florida, 1st edition. ISBN 978-1498727259.

Watanabe, S. (2010). Asymptotic equivalence of Bayes cross validation and widely applicable information criterion in singular learning theory. *Journal of Machine Learning Research*, 11:3571–3594.

Watson, O. and Eaton, J. (2022). *rdhs: API Client and Dataset Management for the Demographic and Health Survey (DHS) Data*. R package version 0.7.6.

Whittle, P. (1963). Stochastic Processes in Several Dimensions. *Bulletin of the International Statistical Institute*, 40:974–994.

Wickham, H., Chang, W., Henry, L., Pedersen, T. L., Takahashi, K., Wilke, C., Woo, K., Yutani, H., and Dunnington, D. (2022a). *ggplot2: Create Elegant Data Visualisations Using the Grammar of Graphics*. R package version 3.4.0.

Wickham, H., Francois, R., Henry, L., and Muller, K. (2022b). *dplyr: A Grammar of Data Manipulation*. R package version 1.0.10.

Wickham, H. and Grolemund, G. (2016). *R for Data Science*. O'Reilly, Sebastopol, California, 1st edition. ISBN 978-1491910399.

Xie, Y. (2022). *tinytex: Helper Functions to Install and Maintain TeX Live, and Compile LaTeX Documents.* R package version 0.43.

Xie, Y., Allaire, J., and Grolemund, G. (2018). *R Markdown: The Definite Guide.* Chapman & Hall/CRC, Boca Raton, Florida, 1st edition. ISBN 978-1138359338.

Xie, Y., Cheng, J., and Tan, X. (2022a). *DT: A Wrapper of the JavaScript Library DataTables.* R package version 0.26.

Xie, Y., Dervieux, C., and Riederer, E. (2022b). *R Markdown Cookbook.* Chapman & Hall/CRC, Boca Raton, Florida.

Zhong, R. and Moraga, P. (2023). Bayesian hierarchical models for the combination of spatially misaligned data: a comparison of melding and downscaler approaches using INLA and SPDE. *Journal of Agricultural, Biological, and Environmental Statistics.* doi: 10.1007/s13253-023-00559-w.

Zhong, R., Ribeiro Amaral, A. V., and Moraga, P. (2023). Spatial data fusion adjusting for preferential sampling using INLA and SPDE. arXiv:2309.03316 [stat.ME].

Index

For Product Safety Concerns and Information please contact our EU
representative GPSR@taylorandfrancis.com
Taylor & Francis Verlag GmbH, Kaufingerstraße 24, 80331 München, Germany

www.ingramcontent.com/pod-product-compliance
Ingram Content Group UK Ltd.
Pitfield, Milton Keynes, MK11 3LW, UK
UKHW021110180425
457613UK00001B/16